GOVERNMENT DIGITAL

The Quest to Regain Public Trust

ALEX BENAY

DUNDURN
TORONTO

Cover design: Laura Boyle
Printer: Webcom

Library and Archives Canada Cataloguing in Publication

Benay, Alex, author
        Government digital : the quest to regain public trust
/ Alex Benay.

Includes bibliographical references.
Issued in print and electronic formats.
ISBN 978-1-4597-4206-2 (softcover).--ISBN 978-1-4597-4207-9
(PDF).--ISBN 978-1-4597-4208-6 (EPUB)

        1. Internet in public administration. I. Title.

JF1525.A8B46 2018   352.3'802854678   C2018-903175-1
                                                    C2018-903176-X

1 2 3 4 5   22 21 20 19 18

   Conseil des Arts   Canada Council         ONTARIO ARTS COUNCIL
            du Canada   for the Arts   CONSEIL DES ARTS DE L'ONTARIO
                                                                                          an Ontario government agency
                                                                                          un organisme du gouvernement de l'Ontario

We acknowledge the support of the **Canada Council for the Arts,** which last year invested $153 million to bring the arts to Canadians throughout the country, and the **Ontario Arts Council** for our publishing program. We also acknowledge the financial support of the **Government of Ontario,** through the **Ontario Book Publishing Tax Credit** and the **Ontario Media Development Corporation,** and the **Government of Canada.**

Nous remercions le **Conseil des arts du Canada** de son soutien. L'an dernier, le Conseil a investi 153 millions de dollars pour mettre de l'art dans la vie des Canadiennes et des Canadiens de tout le pays.

Care has been taken to trace the ownership of copyright material used in this book. The author and the publisher welcome any information enabling them to rectify any references or credits in subsequent editions.

— *J. Kirk Howard, President*

The publisher is not responsible for websites or their content unless they are owned by the publisher.

Printed and bound in Canada.

VISIT US AT

dundurn.com | @dundurnpress | dundurnpress | dundurnpress

Dundurn
3 Church Street, Suite 500
Toronto, Ontario, Canada
M5E 1M2

CONTENTS

INTRODUCTION

THE VERY CONCEPT of government as we know it is under tremendous pressure due to the arrival of exponential technologies such as autonomous vehicles, augmented and virtual reality (AR and VR), digital biology, artificial intelligence (AI), robotics, and biotechnology, to name only a few, in almost every sector. As a result, the value that public services deliver is being challenged in ways never seen before and at a relentless pace that governments often fail to grasp.

Of course, governments have been under siege for quite some time. In fact, the level of public trust has been decreasing for decades, globalization has forced countries to look at policies at both the micro and macro levels in novel ways, and climate change has steadily shifted the focus of economies. While threats to governments are certainly not new, there is a new "kid on the block" where government threats are concerned. The digitization of the planet is shifting industries, human behaviour, and overall life patterns in new ways we have yet to fully experience. Many experts agree that we are entering a fourth industrial age in which the pace of change is exponential — open, digital, and global — rather than linear — closed, analogue, and local. Almost overnight new digital behaviours and exponential technologies are completely transforming entire industries anchored in centuries of tradition.

Banks, for example, are implementing AI in ways that will render entire professions obsolete. JPMorgan recently implemented a new software called Contract Intelligence (COIN) that conducted 360,000 hours of annual legal work in mere seconds.[1] The Chinese e-commerce company Alibaba, the largest retail chain in the world, doesn't own inventory. Facebook, the biggest global media platform, owns no media. Airbnb, the most extensive hotel chain on the planet, owns no real estate.[2] Digital has touched retail, media, accommodations, transportation, and a myriad of other areas in our lives. Yet governments have been slow, and in some cases, totally absent in adopting new digital practices.

This government inaction poses incredible risks, since public institutions are often called upon to regulate the very industries heading toward new digital horizons at a pace governments simply can't keep up with. A good case is driverless cars. Government operations in many countries still rely on "taxi chits" that public officials manually fill out to move from meeting to meeting. How can governments possibly be expected to understand and regulate industries in which digital is at the core if they can't even apply basic technology in their own internal operations? There is a need for a fundamental rebuild of public service if governments are to grapple efficiently with the digital world.

The danger is quite real and by no means an exaggeration. What happens to regulators when most banks do what JPMorgan has done and introduce AI across all their operations? The two sides won't be able to have a real dialogue because the digital divide will be too great. What will happen to revenue collection agencies around the world once more and more private corporations and citizens adopt cryptocurrencies? The role of government as a regulator, even more important, as a trusted source of authority, is in many ways already broken, and digitization will only exacerbate this complex challenge.

This new digital reality hits every single area of government operations like a sledgehammer. Even in policy development, often seen as a core offering of most democracies in the world, digital realities are making existing practices outdated. For example, public services typically conduct policy development through some form of formal consultation. Historically, consultations occur during a specific period of time — there is a beginning and an end to the consultation period — then a policy is written and implemented. In a world driven by instant social interaction on numerous social media platforms, dialogue in the form of government "consultation" can often be seen as a tokenistic approach for engagement with citizens. How does the public service change its model to better react to events such as the Arab Spring, which was mostly coordinated using social media? How do governments adjust to this new reality in which citizens expect instant engagement with their public institutions because they get an instant response in every other facet of their lives? The policy development mechanism isn't digital, nor is it reflective of new emerging global values demanded by an increasing number of digital citizens.

If policy development doesn't reflect new digital realities, it can also be claimed that government services are outdated. Health care, employment insurance, waste management, and other such services are all impacted by the digital, exponential reality. Consequently, regardless of the level of government — national, regional, or municipal — digital has impacted the expectations of citizens and businesses when dealing with government services.

Citizens are used to tracking their online purchase orders and pinpointing in real time which exact truck is delivering their goods. As individuals, we can order food, music, or transportation from the convenience of our mobile devices. Yet often, when it comes to government services, we assume an analogue delivery. In many ways, this divergent public versus private digital service reality is driven by market needs. In any other sector, there is an imperative to be digital in order to meet shifting market demands. If a service or product isn't digital by design in today's world, it simply fails. In government, this imperative is less present, if at all, since there is no competition, and often a "digital-first" approach to service design is non-existent. The thinking in government hasn't evolved yet because the pressure to do so appears to be less apparent than in other sectors.

If these examples of outdated government operating models in light of today's digital reality appear to paint a bleak picture, it is because the situation is dire. Time is of the essence, because there is a growing fundamental digital divide between governments of all levels and every other sector around the globe. But there is hope. Best practices and successes do exist in government digital design and delivery. For instance, in 2010 while working for OpenText, I was involved in a project in which many government leaders adopted a cloud-based, secure social media application designed in months to meet the "new" iPad need during the G20 summit in Toronto. The solution was designed in partnership between Foreign Affairs and International Trade Canada (DFAIT), now called Global Affairs Canada, which managed the summit operations group, and the Canadian Digital Media Network, a non-profit organization designed to increase digital content in Canada.

The objective was to provide governments and their leaders with a safe space to coordinate the summit and develop policy as opposed to using uncontrollable e-mail. Notably, the project also became the digital secretariat for other countries that took turns hosting G20 summits. But user needs and platforms change quickly in the digital world, and if these aren't anticipated and acted upon with timely urgency, yesterday's innovations will die a quick death.

Nonetheless, a secure, mobile-based application was indeed developed and delivered in the public sector years before the app-based economy became a reality or cloud technologies were "in." What this teaches us is that innovation is possible in government but the right conditions are required. While working at OpenText, this project was one of the most challenging of my life: dealing with 20 countries, associations, government officials, national leadership cadres, and a multitude of other factors to deliver a program with cutting-edge technology in less than six months. In the end, we supplied a global first.

The G20 experience shaped my career. I went on to provide application and cloud technologies to some of the world's largest and most complex organizations such as the International Olympic Committee and the Commonwealth Games Federation. I also had the opportunity to take institutions anchored in tradition such as the Canada Agriculture and Food Museum, Canada Aviation and Space Museum, and Canada Science and Technology Museum and re-create them as global platforms for true digital citizenship engagement. Today, as chief information officer (CIO) for the Government of Canada, I bring a "digital-first" reality to an entire national government.

Many of the stories in this book were made possible by the support of tremendous leadership. For example, the G20 summit benefited from the direction of the amazing Peter McGovern, an assistant deputy minister at DFAIT back then, who took no prisoners and didn't fear experimentation. During my career, I have been fortunate enough to receive remarkable backing for such change from the chairman of OpenText to the president of the Treasury Board and the secretary of the Treasury Board Secretariat of Canada. Leadership is key to enabling true, constant transformation, and throughout this book it plays a role in all shapes and forms.

To provide a blueprint for public services of all stripes to adopt more digital practices, this book brings together digital leaders from around the world who bear the scars of enabling digital government transformation. Academics, public-sector leaders, and a few industry captains have all contributed to what it means to be "government digital." Compendiums of lessons learned have been written in the past but never through the lenses of practitioners who have actually driven the change.

The contributions in this book vary from context-setting to more operational perspectives. Carleton University professor Mary Francoli addresses the growing loss of trust in government institutions by citizens, while Iain Klugman, the CEO of Communitech, deals with the relationship of governments and the fourth industrial revolution. Venture capitalists Ray Sharma and Amir Bashir take on fifth-generation wireless (5G) to demonstrate how it will be the backbone of smart cities and smart government. John Baker, the founder and CEO of Desire 2 Learn (D2L), tackles digitization and the future of education, while 18F founding member Hillary Hartley argues why it is imperative for governments of all levels to "work in the open." LinkedIn executive Jennifer Urbanski conducts readers through the pathways of social media to show how it should relate to government activity and performance. Olivia Neal, the former director of standards for the Government Digital Service in the United Kingdom, illustrates what is essential to make services digital by design. Siim Sikkut, the CIO of Estonia, shares his country's story on how it became the model digital nation for the planet that it is today, demonstrating how small states can quickly leapfrog large countries to create national digital governments that lead the way for virtual statehood. In my chapter, I use my own experience with the cultural institution Ingenium to show how digital government can be extraordinarily transformative. Lastly, Salim Ismail, formerly from Singularity University in California and the author of *Exponential Organizations*, provides steps toward transforming the mindset of governments from linear to exponential thinking.

This compilation of insights from experts is a world-first in which readers will be able to dive into what it means to be a digital government from all angles, including education and learning, services, and policy development. It also looks at the issues through national, regional, and local lenses, and reveals why evolving to digital is the most crucial challenge governments face in the next decade. Digital government is not one aspect of a series of changes, or simply an enabler for transforming various facets of government operations; digital *is* government moving forward. Thus, this book provides guidance for public institutions worldwide from people who have led some of this change. The contributors are passionate about digitization and want to enable change in government because it is necessary for all societies to advance. Governments around the globe at all levels must adopt different business models to remain relevant. *Government Digital* will help leaders and practitioners of all kinds to understand that such change is, in fact, quite possible, and more important, *required* to remain relevant in what is now firmly a digital planet.

— Alex Benay

NOTES

1. Hugh Son, "JPMorgan Software Does in Seconds What Took Lawyers 360,000 Hours," *Independent*, February 28, 2017, www.independent.co.uk/news/business/news/jp-morgan -software-lawyers-coin-contract-intelligence-parsing-financial- deals-seconds-legal-working-a7603256.html.
2. Klaus Schwab, *The Fourth Industrial Revolution* (New York: Crown, 2016).

1

Trust in an Era of "Open" and Digital Government

Mary Francoli

MARY FRANCOLI, director of Carleton University's Arthur Kroeger College of Public Affairs and an associate dean in the Faculty of Public Affairs, is an up-and-coming thought leader in open government and overall public-sector digital communications. She has been both an advocate and critic of the Government of Canada's efforts in the Open Government Partnership, a global organization with the goal of increasing global access to public-sector information to promote greater democratic values and economic opportunities. Over the years, we've worked together with Mary on several digital challenges, ranging from the impacts of digital media on heritage management to overall governance of information technology. Because of her wide range and depth, it is appropriate here that she focuses on the continued decline in trust in government and how this deterioration is impacted by the sheer pace of everything digital. For years, governments around the world have been losing the trust of their stakeholders in a digital era in which everyone can now order items from their phones or have an incredible amount of data available at the touch of a button.

However, in many ways, governments remain analogue in their approach and even their very design. Mary investigates the impact digitization is having on this lack of trust and where governments need to focus to win back the trust of their people.

— Alex Benay

Looking back over the past 17 years, it is possible to establish a bank of rhetoric politicians have embraced internationally regarding the potential for digital technology to improve governance and democracy. Former British Prime Minister Tony Blair, for example, has stated: "I believe that the information society can revitalize our democracy ... innovative electronic media is pioneering new ways of involving people of all ages and backgrounds in citizenship through new Internet and digital technology ... that can only strengthen democracy."[1] Former U.S. President Bill Clinton stated that information and communications technologies (ICTs) would "give the American people the Information Age that they deserve — to cut red tape, improve the responsiveness of government toward citizens, and expand opportunities for democratic participation."[2] Later, in 2009, the British Conservatives stated: "Conservatives believe that the collective wisdom of the British people is much greater than that of a bunch of politicians or so-called experts. And new technology now allows us to harness that wisdom like never before."[3] On September 15, 2016, Canadian Prime Minister Justin Trudeau noted on Twitter: "A more open, digital democracy is one that works for more people."[4]

While perhaps well-intentioned, such statements often leave one wondering about the relationship between rhetoric and practice. A lot has been promised, and a lot of hope has been put into digital technology, but the relationship between government and citizens remains strained, and trust in government has reached historic lows in many countries around the world. The purpose of this chapter is to reflect on the tension between the hope for digital technology and practice and to ask *How might digital technology disrupt the current environment of distrust that plagues many democratic countries?*

In addressing this question, the chapter engages with literature from communication and media studies, public administration, and political science, among others. It starts with a look at the current environment of distrust between citizens and governments before turning to an exploration of the hope, or optimism, that many hold for digital technology. This is followed by a discussion of some of the administrative challenges and barriers to the adoption of digital technology that may have contributed to the unrealized promises made by politicians, such as those noted above. It concludes with a look at efforts that point toward a subtle shift that may be under way to allow the democratizing potential of technology to be more fully experienced in a manner that could, over time, work to rebuild trust, as well as strategies for nurturing these efforts.

TRUST IN GOVERNMENT

The concept of trust is multifaceted and emerges in a range of contexts. Interpersonal trust, for example, focuses on the feelings individuals have toward others.[5] Institutional trust, on the other hand, relates to the attitudes and feelings that citizens have toward institutions. As P.K. Blind states, "Institutional trust is generated when citizens appraise public institutions and/or the government and individual political leaders as promise keeping, efficient, fair, and honest."[6] Blind's definition adds nuance and specificity to more generalized definitions, such as Morton Deutsch's, which simply states that trust is the "confidence that one will find what is desired from another rather than what is feared."[7] While this chapter focuses specifically on the issue of

trust between governments and citizens, more in line with the
work of Blind, it is interesting to note that trust in general,
whether interpersonal or institutional, has been declining in recent years.[8]

Many studies have documented the declining levels of trust
citizens have in government. Between 2007 and 2015, trust in
governments across Organisation for Economic Co-operation
and Development (OECD) countries declined by an average of
2 percent from 45 percent to 43 percent.[9] In some countries,
the decline was much sharper. In the United States, for example,
public trust in the federal government declined from just over
70 percent in 1958 to just under 20 percent in 2015.[10] The 2017
Edelman Trust Barometer shows levels of distrust continuing to
slip between 2015 and 2017 in many countries, including Canada,
France, the United States, and the United Kingdom.[11] Here, we
see that it is not just trust in government that is on the decline but
trust among three other institutions, including business, media,
and non-governmental organizations, which has also declined to
such a point that the barometer has declared the current societal
environment to be one where we are witnessing "trust in crisis."[12]

Within the current trust-related crisis, particularly high levels
of distrust are aimed at government officials. On average, only 29
percent of citizens believe government officials to be trustworthy
or credible.[13] Among other things, declining levels of trust are related to issues like corruption and a growing belief that the system
in general is failing. Worry over abuses of power, a disconnect between government and citizens, and inadequate service delivery
are among the other documented causes of the current strained
relationship between governments and citizens.[14]

The primary consequence of the current crisis in trust is a notable democratic deficit where citizens withdraw from a range of
activities that are fundamental to a healthy and representative democracy. Voter turnout for example has declined. In Canada, voter

turnout federally has hovered between 60 and 68 percent since the 1990s.[15] The statistics are notably lower among youth. In the 2011 federal election, only 38.8 percent of those between the ages of 18 and 24 cast a vote. In the 2015 federal election, participation among this group was marginally better at 57.1 percent.[16] The percentage of the population voting in the American presidential elections has been comparable at 60.2 percent in 2016, 58.6 percent in 2012, and 62.2 percent in 2008.[17] As has been the case in Canada, turnout among youth is also lower than the national average. This scenario is not unique to North America. The United Kingdom has seen similar fluctuations with the turnout during general elections falling from a high of 83.9 percent in 1950 to 67.7 percent in 2017 — a number up slightly from the 2001, 2005, 2010, and 2015 general elections.[18] Other forms of political engagement have also declined, remained very low, or stagnated, such as volunteering for a political party, expressing views by contacting a newspaper or politician, and attending and speaking at public meetings.[19]

To be clear, lack of trust is only one of a range of potential reasons for the democratic deficit. Other factors such as age, education, and socio-economic status are all frequently cited as contributing factors. However, feelings of disconnectedness and mistrust can be fluid, stretching across a myriad of factors. As Jon H. Pammett and Lawrence LeDuc (2003) note in a study of declining voter turnout among youth, feelings of disconnectedness from politics and government, lack of information, knowledge or understanding, distrust of system and politicians, as well as cynicism and disillusionment, are all perceived reasons for declining political participation.[20] Building trust is a starting point for an improved democracy. Moreover, the means by which trust is improved may also erode other reasons for poor engagement. Pammett and LeDuc mention improved education, information, and dialogue, along with greater honesty, responsibility, and accountability in politics, as requirements for improved participation.

The Edelman barometer suggests a new model that disrupts traditional notions of influence and authority, and has some of the same suggestions as Pammett and LeDuc. Here, a shift in practice from a top-down model in which influence and authority are held by a select few to a model in which citizens and institutions work together is thought to be necessary to improve trust.[21] While this, in theory, sounds ideal, it begs a very practical and real question of how such a model could work in implementation, and where the limits of that disruption are situated. Here, many turn to and find hope in digital technology.

THE HOPE FOR DIGITAL DEMOCRACY

The idea that digital technology can be the saviour of democracy is not fundamentally new. In the 1980s, for example, techno-optimist John Naisbitt claimed that the disruption of influence and authority called for by those arguing for the need of a new model of democracy, such as the Edelman barometer as mentioned earlier, was possible and likely on the horizon.[22] This potentially over-optimistic claim was founded on the notion that representative democracy was passé in an environment where technology could allow for a system of direct democracy in which individuals could participate directly in the decisions impacting their lives. In many ways, the woes of democracy were boiled down to problems of time, space, and capacity where citizens had no means by which to voice their opinions across diverse geographical and temporal spaces — a problem theoretically solved by evolutions in digital technology.

Other scholars have been more nuanced, pointing to the ways that technology can improve, not supplant, existing governance structures. Some, such as Kenneth L. Hacker and Jan van Dijk, use the term *digital democracy* to describe the dynamic relationship between technology and government. Digital democracy comprises "a collection of attempts to practice democracy without the limits of time, space, and other physical conditions, using ICTs or computer mediated, instead, as an addition [to], not a replacement for traditional 'analogue' political practices."[23] Most often, the benefits of technology discussed in the literature relate to lowered costs of governance, the provision of information, civic engagement, and the quality of policy. Each is related to the other in some way.

Resource efficiency is a commonly cited rationale for turning to digital technology. It enables governments to push information to a large number of people in a timely and inexpensive manner. Online service delivery, also referred to as *e-government*, allows people to complete administrative forms online and to easily pay fees for various programs or services. Here, we can think of things like the ability to file taxes online or to apply for a new or renewed passport online. While these services can be found useful by citizens, they do not constitute the sort of systemic disruption outlined by the Edelman barometer, nor do they explicitly speak to the potential suggestions for improved engagement offered by scholars such as Pammett and LeDuc.

Efficiencies also extend to the provision of information. Governments are the holders of immense collections of information and data. It follows that such information and data are a significant national resource and should, subject to certain carefully identified restrictions, including individual privacy and national security, be readily available to the public. Thinking of information and data in this manner and providing it in a timely and reliable way relates much more directly to ideas of transparency, accountability, reliability, and improving information and knowledge — all factors identified as important to enhance engagement and trust. The potential for the cycle of distrust to be disrupted is much higher than with notions of e-government. Providing information in this way is often referred to as the foundation for *open government*. Here, digital technology is seen as instrumental. It allows for information and data to be published more quickly and cheaply than before and lets it be more widely disseminated. Internet penetration rates are often pointed to as evidence that the majority of people have the potential to access information and data online. In North America the Internet penetration rate is 88.1 percent, in Europe, 80.2 percent, and there are smaller numbers in other parts of the world, most notably in Africa where it sits at 31.2 percent.[24]

High levels of Internet penetration are also thought to open new possibilities for citizen participation. The sort of direct-democracy style of governance championed by John Naisbitt, in which citizens are engaged in all matters of governance, has not been realized, or even seen as a viable or desirable solution to the ailments of contemporary democracy. However, it is widely recognized that technology affords some increased opportunity for civic engagement. As Mark Holzer and Richard W. Schwester note, digital technology can "help to cultivate a governmental landscape in which information is more accessible, and citizens are better able to participate in political and decision-making processes."[25] Increasing citizen participation helps to bridge the disconnect between government and citizens identified by scholars, such as Berman, as a contributor to lack of trust, and by Pammett and LeDuc, as a potential rationale for low political engagement among youth.

An important caveat when it comes to engagement is that it needs to appear authentic. To do so, governments need to clearly outline the parameters of the engagement exercise and they need to provide citizens with the information and tools required to participate. Here, citizens need advance notice that the engagement will be occurring, information on which to engage, an appropriate platform through which to do it, and evidence that their views have been taken into consideration. Coleman and Blumler refer to this as the "listening government." They state that

> effective listening to the public entails first, making sure that there is a meaningful exchange of

views rather than an almost endless succession of atomized positions; second, engaging in debate with the most prevalent, as well as the most forceful, views that emerge from public deliberation; and third, ensuring that citizens understand when and how their ideas will be considered by government, and what sort of expectations they should entertain in relation to feedback and policy influence.[26]

The third point regarding expectations is particularly important. In its absence, it is difficult to appear to be a "listening government," and engagement activities will be perceived as inauthentic, leading to further mistrust and skepticism.

Improved information and opportunity for civic engagement (if conducted properly) together should theoretically lead to the creation of policy and programs that are more responsive and representative of the needs of the public. In general, the potential is great for a more open and efficient government. The notion of openness is vital here, and it is worth taking a step back to explore in greater depth the concept of open government mentioned above.

Once a term used to refer specifically to freedom of information, or access to information, open government today is a more inclusive term, drawing in elements of its predecessors, but also expanding to incorporate citizen engagement. In part, this elaborated definition can be attributed to the Open Government Partnership. This multilateral initiative, started in 2011, pushes governments to make concrete commitments to improving transparency, accountability, and citizen engagement. It has grown in membership from eight countries in 2011 to 75 in 2017, along with 15 subnational-level governments. As a function of membership, each government must sign the "Open Government Declaration," which not only states a commitment to improving both the quality and quantity of information and data but also to engaging citizens in the co-creation of policy and decision-making. Furthermore, it places a great deal of emphasis on digital technology as the mechanism driving many of these improvements. In short, open government constitutes the sort of disruption to political systems referred to by the Edelman barometer. It nurtures the transparency, accountability, responsibility, and dialogue suggested by individuals such as Pammett and LeDuc and sets out a pathway for governments to clearly show they are the sort of listening government discussed by Coleman and Blumler.

CHALLENGES OF DIGITAL

Opening government by putting more and better information and data online and involving citizens more in decision-making might sound like easy solutions to improve feelings of trust, and ideally, to improve political engagement and the democratic deficit. However, they are fraught with complications and are far from easy to implement in practice. John Naisbitt's hopes for democracy were never realized, since systems of representative democracy continued to survive despite the shrinking of time and space caused by digital media. Scholars such as Robert D. Putnam would argue that the digital world has led to a range of distractions and that largely people are not using the technology for the purposes of political engagement.[27] Others who might share Putnam's skepticism, such as Mark Taylor, Essa Saarinen, and Benjamin Barber, point not just to the lack of will among individuals to use technology to engage in politics but also to the contentious relationship between democracy and technology more generally.[28, 29] Democracy takes time, dialogue, and consideration. Technology can be fast and reactionary. Beyond the potentially contradictory relationship between technology and democracy are other, perhaps more practical, challenges, including outdated legal frameworks, costs, and culture within government.

In many countries, the legal frameworks that govern the functions of the public service can also impact the effective acceptance of technology. In many older democracies, legal frameworks tend to predate digital technology entirely, or predate the widespread diffusion of digital media use in society. Canada is a prime example. Its access-to-information legislation was passed in 1983, long before computers and the Internet became widely used, either within or outside government. It was a different communication era when less information was generated. The Access to Information Act was simply not created in or for the digital world that we know today. The act has not been dramatically amended or updated since the 1980s and remains inapt to the digital environment.[30]

The Official Languages Act is another Canadian example. The act, passed in 1969, gives the French and English languages equal status within the Government of Canada. Essentially, this means that all Government of Canada information and communications released to the public should be in both official languages. This was much easier to do at a time when computers were not widely used, when less information and communication were generated, and when there were fewer channels by which to disseminate information to the public. The fast pace of the current environment has made it challenging for the federal government to effectively implement the Official Languages Act and raises a number of complicated questions. For example, do data sets have to be in both official languages to be released to the public? Do tweets or other social media posts need to be in both official languages?

Cost is another factor and has historically been raised as a cause for concern in the media, potentially contributing to the existing environment of distrust in government. For example, in 2010 in the United Kingdom, the media was critical of government spending on websites calling for "the need for new efficiency in the state's digital space."[31] This came after government open data revealed that £12 million was spent on the planning and design of one website while another was costing taxpayers £9.79 per visit.[32] These inefficiencies came to light only two years after the media reported on £2 billion in government IT "blunders": "The cost to the taxpayer of abandoned Whitehall computer projects since 2000 has reached almost £2bn — not including the bill for an online crime reporting site that was cancelled."[33]

Similar stories could be found in the Canadian media a few years later. Several news outlets took aim at the government's website renewal initiative, which has attempted to migrate many disparate departmental and agency websites into one central government site. According to a *National Post* article, the project "has already cost the federal government at least $9.2 million, of which $5.4 million has been paid. The contract, which was awarded to Adobe Corp. in March 2015, was originally valued at $1.54 million."[34]

Technology can be costly, and "blunders," or perhaps "experiments," such as those mentioned above, do not necessarily inspire confidence among a public that is concerned with the effective use of tax revenue. Carefully considered investments need to be made to ensure that hardware and software are available to support digital initiatives and that resources are also provided to maintain and operate technology. A closer dive into many of the critical media articles in the United Kingdom and Canada, such as those noted above, show that in some cases, large sums of money were spent on contractors as there were skills that were missing within government for the planning, design, and implementation of certain technology-centred projects. In an era of distrust and skepticism, governments may be more risk-averse in their technology investments, particularly if there is no guarantee that the technology will evolve and function as planned, as has been the case with the consolidated Government of Canada website. And, even when spending decisions move forward, the speed of technological change makes it next to impossible for governments to stay current with technology.

Processes for procurement and implementation are lengthy and do not necessarily allow governments to be nimble and agile.

While legal frameworks and resource challenges can be hard enough to overcome, perhaps more difficult is changing the attitudes and cultures of those who work in government. In many large bureaucracies, the culture is described as being "a culture of hierarchy, risk aversion and lack of trust."[35] Widespread cultural change takes time, requires effort from high level government officials, and is not easily achieved through the adoption of technology alone. In many cases, widespread technological adoption can contribute to the restrictive cultures of government. As Richard Mulgan notes, "the fact that more public service advice may end up in the public arena places officials under greater pressure to compromise with the truth in the interest of not undermining the credibility of their political masters."[36]

MAKING THE SHIFT

Despite all the challenges and barriers associated with digital technology, there is a lot of evidence that most governments have taken steps to embrace the digital world. The United Nations e-government surveys, for example, tell us how well countries are doing when it comes to embracing technology for e-government and e-participation. Canada has consistently ranked in the top 20 since 2003.[37] The United Kingdom, Australia, New Zealand, Singapore, and Finland are also among the countries that ranked at the top of the 2016 survey. Similarly, the Open Data Barometer shows that Canada ranks second when it comes to how well the government is releasing and using open data for accountability, innovation, and social impact, whereas the United States ranks fourth and the United Kingdom ranks first.[38]

However, the continued decline in trust toward government clearly demonstrates the failure of past digital strategies or efforts.

There may be new, technology-driven opportunities for citizens to participate in government, but those opportunities are either not being realized or are not perceived as genuine opportunities. Thus, the democratic deficit and the issues of trust in government persist. The challenges and barriers discussed above often prevent government from making bolder, more innovative use of technology. As Jeffrey Roy notes:

> Increasingly, then, the public service and elected officials are sending mixed signals about the importance of both public participation and online democracy. The reluctance by the government to entertain bolder reforms appears to be consistent with the cautious, incremental approach undertaken in terms of service delivery reforms.[39]

There is evidence that a subtle shift may be under way with the commitments countries are making under the guise of the Open Government Partnership (OGP). However, in many cases, these commitments are being made with no funding attached to them, which makes it difficult for public servants charged with implementing them to propose ambitious changes or to adopt ambitious proposals suggested by the civil society. This, coupled with a culture of risk aversion, lends only to the typical, incremental change approach discussed by Roy. Over time, this, too, will be at risk of becoming a driver of distrust, instead of trust, as it is intended.

Moving forward, governments will need to be willing to embrace more ambitious change and to pay particular attention to Coleman and Blumler's characteristics of a listening government. This may require resources, the willingness to take risks, and to make mistakes. It may also require innovative solutions to dealing with barriers, such as outdated legal frameworks, which do take time to change. At the same time, citizens will need to give some room for failure or experimentation. Governments can help create this room by carefully managing expectations and providing good information regarding goals, spending, and the rationale for change.

It is clear that digital technology can provide the capacity for a more open and engaged government, but as we are witnessing in many jurisdictions, capacity alone will not overcome distrust between citizens and government, and allow for the democratic benefits of technology to be realized.

*   *   *

ESSENTIAL TAKEAWAYS

Renewing trust in government through digital re-engineering is the premise of this entire book. Through the adoption of genuine attempts by governments around the world to be digital and not necessarily digitize analogue processes, the trend of decaying levels of trust in government can be reversed. This journey won't be easy – in fact, it might prove futile in several different ways – but it has to be done. We're serving digital citizens in an analogue fashion in too many countries around the globe. As we embark on this journey, Mary Francoli presents two key factors to consider on this quest to regain public trust through a digital government agenda:

1.   *Technology as an enabler of better democracy isn't necessarily new.* Research and thinking on this topic dates as far back as the 1980s. However, they were rooted in the "efficiencies-of-government" type of mindset, one that still permeates several public-sector organizations to this day. It's important to note that we must evolve from the "technology-can-save-us-money" approach to a "technology-is

everything-we-do" mentality. This shift is occurring at different rates in different countries.

2. *The shift to digital isn't a technology issue.* Most of the legislation in democratic countries around the world was developed, in many instances, pre-automobile. For successful digital transformation, legislation must reflect the digital realities of today. This entails resetting legislation to enable true digital government to be a core item of every government's digital strategy, whether it's politically or public-service-led.

NOTES

1. Mary Francoli, "E-Participation and Canadian Parliamentarians." In Ari-Veikko Anttiroiko and Matti Mälkiä, eds., *Encyclopedia of Digital Government* (Hershey, PA: IGI Publishing, 2007).
2. J.E.J. Prins, ed., *Designing E-Government: On the Crossroads of Technological Innovation and Institutional Change* (The Hague: Kluwer Law International, 2001).
3. Stephen Coleman and Jay G. Blumler, "The Wisdom of Which Crowd? On the Pathology of a Listening Government," *Political Quarterly* 82, no. 3 (2011): 355–64.
4. Justin Trudeau (@JustinTrudeau), "A more open, digital democracy is one that works for more people. Thanks @ EricSchmidt for the meeting today," Twitter, September 15, 2016, 12:48 p.m., https://twitter.com/justintrudeau/status/776508057119514624.
5. J.B. Rotter, "Interpersonal Trust, Trustworthiness, and Gullibility," *American Psychologist* 35 (1980): 1–7.
6. Peri K. Blind, *Building Trust in Government in the Twenty-First Century: Review of Literature and Emerging Issues* (2006), http://unpan1.un.org/intradoc/groups/public/documents/un/unpan025062.pdf.
7. Morton Deutsch, *The Resolution of Conflict* (New Haven, CT: Yale University Press, 1973).
8. Organisation for Economic Co-operation and Development (OECD), *Trust and Public Policy: How Better Governance Can Help Rebuild Public Trust*. OECD Public Governance Reviews (Paris: OECD Publishing, 2017).
9. Organisation for Economic Co-operation and Development (OECD), "Government at a Glance 2017," www.oecd.org/gov/government-at-a-glance-2017-highlights-en.pdf.
10. "1. Trust in Government: 1958–2015," Pew Research Center, November 23, 2015, www.people-press.org/2015/11

/23/1-trust-in-government-1958-2015.

11. "2017 Edelman Trust Barometer," Edelman, accessed October 6, 2017, www.edelman.com/trust2017.

12. "2017 Edelman Trust Barometer."

13. "2017 Edelman Trust Barometer."

14. Evan M. Berman, "Dealing with Cynical Citizens," *Public Administration Review* 57, no. 2 (1997): 105–12.

15. "Voter Turnout at Federal Elections and Referendums," Elections Canada, accessed October 3, 2017, www.elections.ca/content. aspx?dir=turn&document=index&lang=e&section=ele.

16. Conference Board of Canada, "Voter Turnout" (2017). Accessed October 1, 2017, www.conferenceboard.ca/hcp/ provincial/society/voter-turnout.aspx.

17. U.S. Elections Project (2017), accessed October 1, 2017, at www.electproject.org.

18. U.K. Political Info, "General Election Turnout 1945–2017" (2017), accessed October 10, 2017, www.ukpolitical.info/ Turnout45.htm.

19. Martin Turcotte, *Civic Engagement and Political Participation in Canada* (2015), accessed September 29, 2017, at www. statcan.gc.ca/pub/89-652-x/89-652-x2015006-eng.pdf.

20. Jon H. Pammett and Lawrence LeDuc, "Confronting the Problem of Declining Voter Turnout Among Youth," *Electoral Insight* (July 2003). Accessed September 30, 2017, www.elections.ca/content.aspx?section=res&dir=eim/ issue8&document=p2&lang=e.

21. "2017 Edelman Trust Barometer."

22. John Naisbitt, *Megatrends: Ten New Directions Transforming Our Lives* (New York: Warner Books, 1982).

23. Kenneth L. Hacker and Jan van Dijk, "What Is Digital Democracy?" *Digital Democracy: Issues of Theory and Practice* (Thousand Oaks, CA: SAGE Publications, 2000), 1–9.

24. "Internet Usage Statistics," Internet World Stats, accessed July 30, 2017, www.internetworldstats.com/stats.htm.

25. Mark Holzer and Richard W. Schwester, "Citizen Consultations via Government Web Sites," in *Encyclopedia of Digital Government*, ed. Ari-Veikko Anttiroiko and Matti Mälkiä (Hershey, PA: Idea Group Reference, 2007), 1:163–68.

26. Stephen Coleman and Jay G. Blumler, "The Wisdom of Which Crowd?" 361.

27. Robert D. Putnam, *Bowling Alone: The Collapse and Revival of American Community* (New York: Simon and Schuster, 2000).

28. Mark Taylor and Essa Saarinen, *Imagologies: Media Philosophy* (New York: Routledge, 1996).

29. Benjamin Barber, "The Uncertainty of Digital Politics: Democracy's Uneasy Relationship with Information Technology," *Harvard International Review* 22, no. 1 (2001): 42–47.

30. At the time of writing, an amended access-to-information act was being considered before Parliament. The amended act has been heavily criticized by Canada's information commissioner and civil society for not adequately responding to the deficiencies of the original act.

31. Josh Halliday, "Government Website Costs Revealed," *Guardian*, July 5, 2010, www.theguardian.com/technology/blog/2010/jul/05/government-data-websites.

32. Halliday.

33. Bobbie Johnson and David Hencke, "Not Fit for Purpose: £2bn Cost of Government's IT Blunders," *Guardian*, January 5, 2008, www.theguardian.com/technology/2008/jan/05/computing.egovernment.

34. "Cost of Federal Government's Website Renewal Six Times Over Budget, Halfway Through Project," *National Post*,

August 15, 2016, http://nationalpost.com/news/politics/cost-of-federal-governments-website-renewal-six-times-over-budget-half-way-through-project.

35. "Permission to Fail: Changing the Culture of the Public Service," *Canadian Government Executive*, May 7, 2012, https://canadiangovernmentexecutive.ca/permission-to-fail-changing-the-culture-of-the-public-service.

36. Richard Mulgan, "Truth in Government and the Politicization of Public Service Advice," *Public Administration* 85, no. 3 (2007): 569–86.

37. United Nations, *United Nations E-Government Survey 2016* (New York, United Nations, 2016), http://workspace.unpan.org/sites/Internet/Documents/UNPAN97453.pdf.

38. "Open Data Barometer (2016)," Open Data Barometer, accessed September 25, 2017, http://opendatabarometer.org/?_year=2016&indicator=ODB.

39. Jeffrey Roy, *E-Government in Canada: Transformation for the Digital Age* (Ottawa: University of Ottawa Press, 2006), 128.

2

Shift Happens: Governments and the Fourth Industrial Revolution

Iain Klugman

IAIN KLUGMAN, a serial entrepreneur from Waterloo, Ontario, is the CEO of Communitech, one of the country's top technology organizations. He has had a tremendously varied career, ranging from tourism to education and research, and over the past decade has led an organization that is a key innovation and economic engine in the nation. Iain and I have worked together on and off on multiple projects for years, hosting various forums or conferences, such as Canada 3.0, or recently, attempting to increase the connection between technology incubators and the Government of Canada, the largest buyer of technology in the country. Iain addresses the massive shifts we are facing in what has been termed the fourth industrial revolution and how it impacts the world's governments.

– Alex Benay

All over the world there is a distinct sense of the ground moving beneath our feet. We witness many forms of change: disturbing weather patterns, xenophobic protectionism, collapsing states, mass human migration, denial of science, disruption of established industries, distrust of authority, and more.

At the same time we see a radical ramping up of technology's long-standing promises. Data and computing power are growing dramatically. Algorithms are becoming increasingly sophisticated and powerful. Month after month new systems, business models, and capabilities surface and then explode onto the market, offering radical new solutions but also threatening to wipe out huge organizations, throw countless people out of work, and expose personal data to the risk of abuse.

You could say that "shift" happens; in fact, it is happening now and has been for some time. A growing number of experts refer to this radical change as the "fourth industrial revolution." Others see it as the logical conclusion of the digital revolution, or perhaps, the beginning of the second quantum revolution. Whatever we call it, entrepreneurs, innovators, and those who think seriously about the future are now beginning to see how quickly and in what direction the tectonic plates are moving.

It is complicated, to be sure, but we have a duty to reach out to our partners in government and civil society to make sure that Canada is prepared for the next great ride. One thing is certain: we'll either catch this competitive wave or be swamped by it.

*   *   *

World Economic Forum founder Klaus Schwab used the 2016 Davos Summit as a wake-up call. He urged corporate and political leaders to get their heads around the fact that the fourth industrial revolution has not only begun but will alter the world in ways we have never before seen.

According to Schwab, the first three industrial revolutions set the stage: the first (1760s to the 1840s) was all about the construction of railways, the commercialization of the steam engine, and the advent of mechanical production. The second straddled the late 19th and early 20th centuries and brought us electricity and mass production. The third, which dates back to the early 1960s, is the one that really teed up the shift happening now: the emergence of semiconductors, computers, and digital networks.

As Schwab and others point out, the drastic acceleration of computing technology means innovation is not only occurring at an accelerating rate but it's also cutting across previously separate sectors and inflicting massive change on systems we've barely thought about — notably the structures of government.[1]

The forum's current list of key fourth-industrial-revolution technologies includes computing capabilities, storage, and access; big data; digital health; the digitization of matter; the Internet of Things (IoT); blockchain; and wearable Internet. In my view, three additional transformative technologies are still in our labs and in the infancy of application: quantum information processing, sensing, and communications; machine learning; and advanced materials.

Any one of these will be radically transformational all on its own. Some have compared the broad impact of machine learning, for example, to the introduction of electricity in the early 20th century — a technology that changes everything from policing, to medicine, to security. We've all grown up with fixed ideas about the role of automobiles in our lives, but the eventual commercialization of autonomous vehicles (AVs) will upend that familiar image. Robots will help us navigate our lives. Dramatic gains in clean energy point to solutions for global warming.

Then there's blockchain. Most people outside the tech bubble are only vaguely familiar with the term, if at all. Yet this one technology, originally designed to allow bitcoins to circulate securely, will eventually upend any system that involves trusted distributed computation. Blockchain is attracting armies of software engineers and developers, and yet most of us are barely aware of its existence, much less its potential.

Taken together, this torrent of change often feels as if we're perched in the middle of a rapidly moving river on slippery stones that are being shunted around by the current.

"In its scale, scope and complexity," Schwab writes, "what I consider to be the Fourth Industrial Revolution is unlike anything humankind has ever experienced before."[2] Or as Kevin Lynch, the former Clerk of the Privy Council of Canada and vice-chair of BMO Financial Group, says: "What is unique today is the combination of scale and scope of these technologies, their interconnectedness, and the speed of their adaptation."[3]

These developments are playing out against a stormy global environment. International energy markets are in flux, supply chains are dauntingly complicated, and in most industrialized nations, societies are aging, which raises tough questions about who will pay for the extended retirement of so many people.

All this change will be further amplified by the fact that the technologies of the fourth industrial revolution all have the potential to disrupt seemingly stable industries and institutions. Shift will happen everywhere, from routine shop-floor work to the tasks performed by professionals such as doctors, lawyers, and accountants.

Large companies will continue to find themselves toppled by fleet-footed start-ups that begin with little capital and few hard assets. As well, ambitious non-traditional players will elbow into established industries (e.g., the push of Google, Apple, and Uber into automotive manufacturing in order to develop self-driving vehicles).

These phenomena are playing out against a backdrop of rising concern about technology companies themselves — from their stewardship of users' personal data, to the lack of gender and racial diversity within their ranks, to the concentration of power in the hands of a few large companies, all of which have fuelled calls for government regulation.

It's difficult to predict a lot of what's to come, but this much is certain: we will see dizzying shifts in labour markets and will become part of a society that is forced to think through the ethical, professional, and legal implications of technologies that can outperform us.

Historically, such periods of technology-driven upheaval have also brought on productivity gains, economic growth, improvements in quality of life, and increases in longevity/health. There's no reason to believe that the fourth industrial revolution, like the three that preceded it, will fail to deliver these long-term benefits, especially in a world where billions of people still aren't connected to an electrical grid.

While technological development will eliminate some forms of work, the productivity gains will create new wealth, new investment, and new jobs. As Schwab points out, 90 percent of the U.S. labour force worked on the land at the beginning of the 20th century, whereas today that number has fallen to 2 percent, thanks to the mechanization of agriculture, mining, and forestry.

But periods of great disruption have also fostered waves of political and social instability. The new global normal, Kevin Lynch says, is a world filled with ever-mutating forms of risk: cyber attacks/espionage, terrorist threats, failed states/institutions, loss of confidence in government, mass involuntary migrations, protectionism, and the rise of nationalist or xenophobic populism.[4]

For any government, industry, and civil society, the fourth industrial revolution must become a top priority before the shift hits the fan. These are the big questions that should keep everyone up at night.

*   *   *

One way or another, this storm will come to every nation, and it's really up to citizens to determine the extent to which their country will benefit as a society.

The gains from the game-changing innovations looming on the horizon are extraordinary. "Digital technologies are doing for brain power what the steam engine and related technologies did for human muscle power during the first industrial revolution," observe MIT Sloan management professors Erik Brynjolfsson and Andrew McAfee, co-authors of *The Second Machine Age*.[5]

Shared or privately owned autonomous electric vehicles will allow us to reduce our dependence on fossil fuels, improve road safety, and allow individuals or families to drastically reduce the amount they spend annually on car ownership.

Machine learning will disrupt banking, law enforcement, law, and almost every other data-dependent sector. This technology will have game-changing implications for health care, wellness, and prevention. Algorithms that read gene sequences can already be used to bio-print tissue designed specifically for a patient. Smart watches with subtle heart sensors will use artificial intelligence (AI) to detect heartbeat abnormalities and warn wearers that they should seek out medical attention.

IoT, in turn, holds the promise of highly integrated but dispersed networks of "smart factories" where individual pieces of equipment are connected to sensors and the Internet, allowing faster and more efficient production. General Electric is already pushing forward with this transformation, setting a goal of connecting 75 of its approximately 500 factories by 2016.[6]

The emergence of new high-strength, low-weight materials will change everything from manufacturing to architecture. Three-dimensional (3D) printing is poised to revolutionize the mass-production paradigm, creating new markets for inexpensive but highly customized goods. These will eventually include microchips and ultimately human tissue. Some experts believe that within a decade, surgeons will be transplanting fabricated body parts rather than relying on donors.

Scientists equipped with quantum computers will be able to model the universe and anything in it using tools that employ its most fundamental laws. Deep mysteries with huge practical significance in fields from superconductivity to climate dynamics will be resolved.

Yet some of these technologies, especially those that automate routine tasks, may trigger job losses. A recent McKinsey & Company study predicts that almost half the time workers spend on their jobs can already be replaced with existing technologies.[7]

Some banking executives have predicted that online "chatbots" that rely on AI-driven natural-language processing will lead to branch closures and call-centre personnel layoffs. Other industries that rely on call centres — insurance companies, telecommunications giants, et cetera — will follow suit.

The advent of adaptive robots equipped with state-of-the-art sensors may also eliminate both routine and complex work. Some industries are especially conducive to automation because of recent efforts to simplify the production and distribution process as part of offshoring strategies.[8]

Autonomous vehicles, in turn, could eliminate all sorts of driving-related jobs, including truck drivers, couriers and transit operators. The AV revolution isn't just about four wheels, either. It will eventually include drones, boats, and perhaps even underwater vehicles.[9]

The job-loss scenarios have prompted gloomy warnings from leading American economists such as Joseph Stiglitz and Larry Summers. Tesla founder Elon Musk wants governments and civil society actors to ensure that machine-learning systems are deployed — ethically. Microsoft founder Bill Gates wants governments to find ways to tax robots to compensate for mass worker displacement.

Most economists predict mid- and long-term productivity gains, coupled with the emergence of new industries and forms of work. Still, the transition period we're entering is extremely fluid, and frankly, scary. We all need that next paycheque, a safe place to sleep, and money for groceries. If fourth industrial revolution technologies lead to chronic unemployment, will political and social unrest follow?

We only have to look at the opioid crisis and the sense of disenfranchisement of un- or underemployed American blue-collar workers to understand why we're living in a period of rising nationalism, xenophobia, and protectionism. Public opinion polls in many industrialized nations show rising distrust and disillusionment with governments, public institutions, and even democracy itself.[10] Meanwhile, massive social media platforms have been implicated in facilitating the spread of propaganda and "fake news," and in allowing their users' personal data to be accessed by political campaign operatives. If these technologies produce waves of layoffs, uncertainty, and political polarization, social pressures could mount, bringing greater income inequality and anger, as well as a backlash against the policies and programs meant to foster innovation.

*   *   *

Canada is an example of a country that is well prepared for the fourth industrial revolution. Demand for its natural resources and its excellence in agriculture and mass manufacturing has kept its economy strong and growing for decades. Canada's financial institutions are widely acknowledged for their stability. Canada's debt-to-GDP ratios are low by international standards, and the country likes to trade with the rest of the world. It's a nation that embraces immigration and keeps inequality in check. And unlike the United States and the United Kingdom, Canada's politics and public institutions aren't under attack. As a 2017 ranking on the Social Progress Index showed, the country is sixth out of the 133 nations measured and ahead of all other members in the G7. The Reputation Institute scored Canada first on its 2017 scorecard ahead of Switzerland, the Scandinavian nations, Australia, and New Zealand.

Over the past two decades, Canada has leveraged its assets to earn a seat among global leaders preparing for the shift to the fourth industrial revolution. But there is still more that it can do to better thrive in a world experiencing the fourth industrial revolution. Complacency is always risky, but never more than now.

For generations Canada has grappled with the two-headed problem of stubbornly low productivity and its status as a so-called branch-plant economy. Despite the billions of dollars spent on advanced research, the federal government admits that its "innovation metrics" lag behind those of other industrialized countries.

Canada isn't lacking in smart, entrepreneurial people. It ranks high on international measures because of attractiveness to foreign visa students (Canada is the eighth most popular destination) and highly educated immigrants. Its leading academic institutions are filled with globally renowned researchers.

But among Canadian high school students, math and science scores on the Organisation of Economic Co-operation and Development (OECD) Programme for International Student Assessment (PISA) rankings have been dropping for over a decade. Many newcomers with advanced technical degrees end up in jobs that don't tap their skills. And Canada is in the bottom fifth of OECD countries in terms of adults with the training required for jobs in the information and communications technology (ICT) sector, which is one of the key drivers of the fourth industrial revolution.

Another important aspect to pushing forward innovation is the funding and commercialization of fundamental and applied research. Canada spends more per capita on higher education than any other OECD nation. Canadian research is frequently cited internationally, and it's gaining a global reputation in strategically vital fields such as machine learning, robotics, life sciences, quantum information, and biomedical engineering.

But if countries truly want to brace themselves for the giant shift sweeping around the world, conducting research and producing talented people are only part of the puzzle. Today a larger problem is finding better ways to allow entrepreneurs to commercialize the technology that percolates up from university labs. The ability to incubate and grow globally competitive companies is central to any country's ability to survive the shifts that are coming at them.

*   *   *

So how can any country looking to brace itself for the fourth industrial revolution get its shift together? Before addressing that question from a few directions, let's start with high-level principles:

1. This is no time to let up. We have to keep investing as heavily as possible in the technologies, people, and transition strategies required by the fourth industrial revolution.
2. Policy-makers, innovators, and researchers all have to think about partnerships differently and make new friends. We all have to stop talking to the same people.
3. Politicians and officials have to learn to leverage ecosystems in the same way that entrepreneurs do.
4. We all must come to understand the difference between innovation and digitization. They're not the same.
5. Governments have no choice now but to run fast. Jurisdictionally, competitiveness is key. People, capital, and companies have many options, and they can pass us by.
6. Governments should also view regulation as a point of differentiation and source of competitive edge.
7. In the era of technologies such as blockchain, centralized planning simply no longer makes sense.
8. Elected officials must draw inspiration from entrepreneurs in order to engage and re-engage citizens, because if they're not part of these big conversations, shift will never happen.

The above list is long, but how should a government oper-
ationalize these principles to avoid being overwhelmed by the
changes swirling across the globe?

According to Thomas Philbeck, the World Economic Forum's
global leadership fellow, policy-makers, innovators, and corporate leaders can begin with a broad framework for building
highly flexible institutions capable of not only surviving shift
but emerging stronger.

- **Think in terms of systems, not technologies.** "Individual
  technologies are interesting, but it is their systemic impact that
  matters," Philbeck says.
- **Empower, not determine.** "Technologies are not forces," he
  says. "We have the ability to shape them and decide on how
  they are applied."
- **Envision a future by design, not default.** "[F]ailure to pay attention to critical governance questions in consideration of the
  Fourth Industrial Revolution means societies are likely to allow
  undemocratic, random, and potentially malicious forces to shape
  the future of technological systems and their impact on people."
- **Regard values as a feature, not a bug.** "They can be embedded
  in technological systems through organizational cultures and by
  prioritizing particular outcomes and rewarding behaviors."[11]

Now let's take it down to ground level. From the perspective
of business, the shifts brought on by the fourth industrial revolution will drastically alter customer expectations about an impos-
sibly wide range of products and services. They'll be able to use
AI to earn better returns on a robo-adviser-managed portfolio.
Their dermatologist will be able to distinguish between a mole
and a pre-cancerous lesion with a camera. Personalized robots
will anticipate their daily tasks. Consequently, the exploitation
of data and digital innovation will become central features of

all product development processes.[12] Companies without an AI strategy and a big-data strategy are doomed.

Governments, in turn, must create the conditions that enable firms to bring these kinds of innovations to their customers. But they also have to internalize the lessons of shift acknowledging the chasm between their own limitations and the dynamism of this emerging world.

Shift-minded policy-makers should borrow ideas and technologies from innovation-oriented companies to make government services as flexible, fast, and competitive as those offered by the private sector. They also have to develop new ideas for providing a safety net for those caught in the dislocation of shift.

So far, many governments have embraced the rhetoric of innovation, and they're investing tax dollars to that end. But more needs to happen for them to truly start changing how they operate and provide better services to citizens.

It begins with trust. While transformative innovation comes from scientists, entrepreneurs, and visionary investors, they play by the rules the government sets. Currently, many industrialized nations are experiencing a sharp decline in public confidence.[13] Anti-government populism has its roots in forces that are flourishing around the world. Still, our governments have to rebuild credibility before they can propose policies that will enable innovators to compete in a shift-driven world.

The regulatory systems of most countries were caught flat-footed by the emergence of sharing services such as Airbnb and Uber that have won enormous consumer buy-in, although not without creating new concerns of their own. And many thorny regulatory challenges still loom with respect to autonomous-vehicle liability, medical ethics, and advanced robotics. Regulatory systems have to become far more adaptable and responsive, not to blindly clear away all impediments to new innovation but rather to narrow the gap between policy and technology.

While regulators look for ways to enable instead of obstruct-ing, ministries and public agencies should be strengthening the in-novation ecosystem with their procurement dollars. Governments should modernize their operations so residents can interact with public services in ways that at least echo the fast-paced digitization that's shifting the rest of the economy. We badly need to close the gap between what consumers can get in the marketplace and what they expect from their governments.

Government and private sector procurement of cutting-edge technology can also provide a boost to emerging innovation firms competing in these transformational industries. Recent efforts, such as Canada's ScaleUP Ventures, are connecting start-ups with large corporate and public-sector customers. The CAD$100 million venture fund, which closed in September 2017, includes commit-ments by some provincial governments as well as large firms in auto parts, banking, and telecommunications.

These initiatives all speak to one of the most important fea-tures of the fourth industrial revolution; advances in the develop-ment and commercialization of these technologies must incubate in ecosystems and clusters that are capable of attracting a critical mass of talent, capital, and customers. For example, Canada has several well-established clusters — Southern Ontario's automotive sector; the concentration of tech firms in the Toronto-Waterloo Corridor; and Montreal's aerospace, telecom, and system integra-tion communities. The federal supercluster program is an import-ant move to build on national strength and to help Canada catch up in the international cluster policy arena.

A 2016 Canadian federal government report on an inclusive in-novation agenda points to other world-leading cluster/innovation programs such as the ones in the United States, the Netherlands, Ireland, Australia, and Germany. In the United Kingdom, a net-work of highly regarded Catapult Centres fund and commercialize late-stage research and development in several key areas such as renewable energy, future cities, smart manufacturing, and medical discovery. The goal of the Catapult program is to transform the United Kingdom's innovation capacity.

Why are clusters so key to survival through the fourth industrial revolution? As Schwab and others predict, the transformation triggered by these new technologies will touch every part of our society. New businesses and industries will spring up; others will die. Some of our most stable companies and sectors will survive, but they'll emerge altered, perhaps beyond recognition. Shift will happen.

It's crucial for innovators and investors to be at the centre of all this global upheaval, and just as important for governments to set the conditions that will allow them to compete. If they do, everyone will realize benefits, not just improved productivity and growth but the cultivation of globally oriented tech companies that serve everyone around the world.

For example, a 2017 McKinsey study concluded that transforming the Toronto-Waterloo tech corridor in Canada into a global innovation hub "has the potential to deliver a $50-billion increase in direct equity value, $17.5 billion in direct annual GDP, and more than 170,000 high-quality jobs by 2025." Toronto-Waterloo tech firms employ more than 200,000 workers (2016 figures), enjoy access to major financial services, and can draw on graduates and research from several major universities.[14]

With public policy investments meant to transform this super-cluster into one that rivals Silicon Valley or Tel Aviv/Haifa, governments can create a virtuous circle that creates great jobs, attracts entrepreneurs, and produces an economic updraft. I would argue that the nations that can incubate innovation-driven superclusters using a combination of smart regulation, smart procurement, and smart research policy will be best positioned to survive the coming shifts.

As Schwab writes, "We stand on the brink of a technological revolution that will fundamentally alter the way we live, work, and relate to one another. In its scale, scope, and complexity, the transformation will be unlike anything humankind has experienced before. We do not yet know just how it will unfold, but one thing is clear: the response to it must be integrated and comprehensive, involving all stakeholders of the global polity, from the public and private sectors to academia and civil society."[15]

* * *

ESSENTIAL TAKEAWAYS

Iain Klugman describes the changes governments are facing in every nation. The possibility exists for mass job displacement on a global scale for countries that aren't invested in the fourth industrial revolution. Jobs, people, skills, and economies will shift to countries that have the right conditions to accommodate this new reality, one in which exponential growth and disruption are the new norms. Here are some key takeaways from his work:

1. *Policy-makers in the public sector aren't necessarily ready for this shift.* Many governments still see digital as a back-office function when in reality the world has switched to being digital, which is enabling the current fourth industrial revolution. Policy developers at all levels of government around the world often neglect this reality and continue to operate in a pre-digital age to their own detriment.

2. *The world of distinctive sectors that occasionally collide is now over.* Everything must be seen as cross-sector in approach. Medical research is based on big data, and data scientists must now depend on the rapidity of governments to adjust their regulatory frameworks to enable different types of research. We live in an interconnected society, something out-

lined by Ray Sharma and Amir Bashir in Chapter 3, which is stimulating rapid cross-sector growth and change as never before. Governments and their civil services must adjust their engagement methods to be more relevant.

3. *In many cases, governments around the globe aren't equipped with the proper skills, experience, and know-how to adjust to this new reality.* The fact that we're transcending to a digital reality that's as prominent, or in many cases, more prominent than our physical realities, is something new to governments.

NOTES

1. Klaus Schwab, *The Fourth Industrial Revolution* (New York: Grove, 2016), 3.
2. Schwab, 1.
3. Telephone interview with Kevin Lynch, October 22, 2017.
4. Kevin Lynch.
5. Erik Brynjolfsson and Andrew McAfee, *The Second Machine Age: Work, Progress, and Prosperity in a Time of Brilliant Technologies* (New York: W.W. Norton, 2014).
6. Christopher Mims, "How U.S. Manufacturing Is About to Get Smarter," *Wall Street Journal*, November 13, 2016, www.wsj.com/articles/how-u-s-manufacturing-is-about-to-get-smarter-1479068425.
7. Christopher Mims, "Technology vs. the Middle Class," *Wall Street Journal*, January 22, 2017, www.wsj.com/articles/technology-vs-the-middle-class-1485107698.
8. Schwab, *The Fourth Industrial Revolution*, 40.
9. Schwab, 15.
10. Amanda Taub, "How Stable Are Democracies? 'Warning Signs are Flashing Red,'" *New York Times*, November 29, 2016, www.nytimes.com/2016/11/29/world/americas/western-liberal-democracy.html.
11. "The Fourth Industrial Revolution: Proceedings of a Workshop — in Brief," National Academies of Sciences, Engineering, and Medicine, March 2017, www.ncbi.nlm.nih.gov/books/NBK424911/pdf/Bookshelf_NBK424911.pdf.
12. Schwab, *The Fourth Industrial Revolution*, 53.
13. While Canada's political landscape is not nearly as polarized as in the United States, the Edelman Trust Barometer in its 2017 report found that for the first time since the survey has been carried out, more than half of all Canadians expressed distrust for key institutions such as the government, media, business, and NGOs. The survey also showed a growing gap

between the "informed public" and the population at large. "Canada at Populism, Trust Crisis Tipping Point," *Cision*, February 14, 2017, www.newswire.ca/news-releases/ breaking-canada-at-populism-trust-crisis-tipping-point-613678663.html.

14. "Tech North: Building Canada's First Technology Supercluster," *NEXT Canada*, December 14, 2016, www.nextcanada. com/images/TechNorth-McKinsey-Report.pdf.

15. Klaus Schwab, "The Fourth Industrial Revolution: What It Means and How to Respond," *Foreign Affairs*, December 12, 2015, www.foreignaffairs.com/articles/2015-12-12/fourth-industrial-revolution.

3

Fifth-Generation Wireless: The Backbone Behind Smart Cities and Smart Government

Ray Sharma and Amir Bashir

I MET RAY SHARMA in 2014 during the inaugural launch of the Canadian Open Data Experience (CODE), the country's largest hackathon at the time, which was designed to leverage federal data to create solutions and new applications. Ray and his business partner, Amir Bashir, are two of Canada's most prolific venture capitalists, with deep ties in Toronto, London, and Silicon Valley, yet remain very humble in their approach. Ray's passion for technology and the social good is what keeps me coming back to him. He is a successful and proud entrepreneur who cares about public policy, and personally, he is a friend and trusted source on many topics. His contribution to this project is a very technical one. But, then again, networks are complex, and we take them for granted as our signals are beamed all over the planet while 99 percent of us never stop to truly think about how they actually work. The importance of this chapter is how it showcases the possibility of accelerating human technological developments by a factor of 100. Stop and consider that for a moment. We are on the verge of entering an exponential growth period around the world and will soon have the potential to increase productivity on an unprecedented scale. This ability will have a tremendous impact on everything from policy to service design and the overall structure of governments everywhere.

– Alex Benay

Fifth generation (5G) is so superior to any prior mobile network experience that it promises to revolutionize the software application experience for citizens and institutions on a supremely broad basis. In fact, the buzzwords of the moment such as Internet of Things (IoT), autonomous vehicles (AVs), artificial intelligence (AI), and beyond are all pinning their success on a network that will require a dramatic evolution to support a new generation of application software.

After exploring the technical nature of 5G, it will become clear that there are implications for software architecture, development, and user interface, and that all industries will be impacted, from the mass-market consumer to the most basic of industrial applications. Governments will be similarly or more impacted than any industry. However, there are some unique distinctions. The first is that 5G will enable smart city and smart government design. The second is that governments are the managers and largely the owners of a wide variety of spectrum, or the air frequency used to transmit data between devices. The essential nature of 5G is that it can leverage unused spectrum to deliver its tremendous performance. Governments are 5G insiders with the potential to play central roles in 5G and its overall economic effect.

The headlines about 5G will focus on the massive increase in download speeds. 5G promises to be 100 times faster than fourth generation (4G) and 25,000 times more rapid than third generation (3G)![1] Mobile users will now be able to experience gigabyte-per-second download speeds, which will allow video to become the dominant form of traffic on the Internet and comprise over 75 percent of all Web business by 2020. Enabled by gigabyte-per-second download speeds, high-definition video conferencing and next-generation concepts such as virtual reality and augmented reality will become mainstream. The public sector will be able to bring town halls online so that large-scale Internet meetings will become seamless and fluid, making possible audiences far larger than current in-person ones.

While we will explore the implications of a substantial boost in speed, the more significant impact is in another feature of the new 5G network. The 5G network calls for a reduction in latency, or the amount of lag time from the trigger to the response of an action, from 100 milliseconds to less than one. Latency is grossly underestimated, since it's a less precise variable than raw download speed, but it's the massive reduction in latency that will have the greatest impact on application development.

Imagine sitting in an audience listening to a lecture. The network latency of 4G is approximately the same as the delay between the brain and the audio registering in the ears. The latency of 5G is difficult to express in relatable terms, since it is 1 percent of the lag of 4G, roughly one-tenth of the delay of the visual system and the brain, which in itself is one-tenth of the delay of audio.

The bottom line is that 5G will enable a new era of machine-to-machine (M2M) interaction. Before a user finishes opening an application, the machines will have spoken to each other so many times that it will require a new interaction experience with the users, which will have a profound impact on concepts such as autonomous driving. For all of that to become a reality, vehicles will need to communicate with public infrastructure such as roads and traffic lights, which will have to become "smart themselves." This M2M interaction with real lives at stake only becomes feasible with one-millisecond latency, and municipalities will be the drivers behind modernizing cities to meet this need.

We are on the verge of historic opportunities for governments and innovation. The former will have a larger-than-ever influence over the next wave of innovation as 5G facilitates the creation of the first "smart cities," where billions of devices are connected, where open and accessible data is encouraged, and where M2M interactions are expected between both the private and public sectors. The industries of tomorrow will be more tightly knit than ever, and governments will be responsible for both investing in new digital public infrastructure as well as distributing new spectrum in a fair and equitable way.

The next-generation wireless network technology is horribly misnamed as 5G because it is such a significant improvement in experience that it will usher in a new era of applications.

SMARTPHONES PROMPT SMART NETWORKS

Over 27 million Canadians are mobile users today, 21 million of whom are using smartphones like the iPhone.[2] These smartphones are portable supercomputers that allow us to communicate, work, and entertain ourselves anytime, anywhere. In fact, the A11 Bionic chip that powers the latest iPhone X can perform a stunning 600 billion operations per second.[3] With millions of apps available to download, 46 percent of respondents in a Pew Research study say they can't live without their smartphones.[4]

This mass-market addiction comes at a cost. Users who could get by with a 500-megabyte data plan five years ago now find themselves upgrading to multi-gigabyte plans. With video and interactive elements such as augmented and virtual reality, data usage is expanding faster than most expectations.[5,6]

To support this massive growth, telecommunication operators must continuously upgrade their networks to ensure capacity, speed, reliability, and availability. Approximately every eight to 10 years, these operators make a generational upgrade to their networks (e.g., from 3G to 4G), which often has profound implications for the kind of devices and applications that reach consumers. It is this generational shift that led to the creation of the BlackBerry in the early 2000s as well as to the second generational shift that gave birth to today's app economy.

Apple and Research in Motion (now renamed BlackBerry after its famous device) deserve much of the credit for laying the foundations of the smartphone economy. Research in Motion was the architect of the smartphone industry, while Apple was the primary developer of the app business. The two are the most innovative firms in technology history but also represent the prodigious creation and destruction of value in the smartphone world.

The history of the app economy can be seen as four evolutionary waves. The prequel was the core communication apps within the BlackBerry. Phase 1.0 introduced consumer apps such as games, social networks, and over-the-top (OTT) messaging. Phase 2.0 was characterized by the broad-based adoption of software to serve the needs of organizations rather than individuals so that apps became productivity tools and tablets were now legitimate computing platforms. Today, 10 years later, Phase 3.0, largely to be driven by 5G, will usher in the industrial and M2M era.

The industrial era of applications will present a new set of challenges. Networks will need to support hyper-dense user numbers. For example, the City of Toronto will have to support up to one billion network-connected devices. And because we are starting to connect everything from parking meters in cities to soil sensors in farms, vastly improved battery life will be crucial.

While governments and businesses will activate programs to leverage 5G for the perceived benefits in regulation (monitoring) and administration (cost savings), longer term benefits are more substantial. The smartphone is positioned as the personal connection point to the Internet, but with emerging industrial apps and IoT, 5G will enable an entirely new experience with the Web.

Some of this evolution will be hidden behind machines or sensors in remote fields of endeavour, but users will ultimately be impacted as a new human-computer paradigm is created.

FIFTH GENERATION: A COMPLETELY NEW MOBILE EXPERIENCE

We know that 5G aims to be 100 times faster than 4G networks and 25,000 times quicker than 3G networks. After the shock of such an improvement diminishes, it is easy to underestimate 5G as just an opportunity to download movies more rapidly.

To understand why 5G is a completely new mobile experience, we'll look in detail at what it will deliver. The core attributes to be discussed are the following:

1. Gigabyte-per-second download speeds
2. Latency (delay time) improvements that redefine software experience
3. Massive Internet of Things (IoT)

GIGABYTE-PER-SECOND DOWNLOAD SPEEDS

Video has taken over the Internet and its growth shows no signs of slowing down. In 2016, 57 percent of global mobile data traffic was video, and by 2021 this is expected to reach 77 percent.[7] Reports analyzing North American data reveal that YouTube alone has triple the data traffic of all other Web surfing combined.

The 5G network will address this demand by bringing faster networks and higher capacity online, so what implications does that have for future applications? It will make the biggest impact on real-time video, especially augmented/virtual reality and videoconferencing.

Watching videos and movies is already a pretty good experience on most 4G networks, and due to the efficiencies of streaming, it's rare that an entire file needs to be downloaded instantly. Multi-gigabyte data rates will allow streaming of movies in high resolution, while terabyte-level capacity will make it possible for people to do so at the same time. Netflix, YouTube, Amazon Prime, and Hulu will continue to proliferate as it becomes normal to watch videos on the go. As a testament to this, Netflix and YouTube represent over half of all fixed-access data traffic at peak hours in North America, and we are likely to see the same outcome in mobile over time.

Rich video content doesn't end at streaming movies. Governments and large corporations will be able to conduct high-definition videoconferencing with such ease and low cost that the traditional concepts of the office will be challenged. Each 5G device will be capable of videoconferencing quality that previously required dedicated boardrooms and hundreds of thousands of dollars in information technology (IT) investment. It will then be possible for governments to hold town-hall meetings remotely using real-time high-definition video conferencing. In cities this could improve attendance for senior citizens and those with accessibility needs who are currently restricted in their ability to appear in person. In rural communities, this could become the dominant form of public communication, especially in regions where citizens live far apart or harsh weather makes transportation unsafe.

Even within governments themselves, high-definition video conferencing further reduces the need for physical presence, which will undoubtedly greatly influence the way we work in the future. How will 5G transform office design? How will meeting rooms change? Large corporations are already moving toward hotel-like situations where employees no longer have assigned desks but are encouraged to be mobile. This means more meetings are happening online, and with 5G, the experience will become even more seamless. Very few organizations in the world own more physical assets than governments. With 5G, all levels of government have a remarkable opportunity to lead in the optimization of managing human resources and physical assets.

Open data is a government-led initiative to release the massive data reserves of government institutions to the public. A popular example of open data employed daily is for the weather. But this is only a very small part of the immense amounts of data collected by governments. With 5G we will see a flood of data unleashed, and with it, the ability to incorporate entire data sets into applications that will either improve the productivity of public institutions or forge new ways for citizens to engage with their governments.

For example, in June 2017, the Government of Canada enlisted ThinkData Works, a Canadian technology company, to analyze trapped data from different departments to find out who the government was awarding IT contracts to. In large organizations, it's incredibly difficult to access information from across departments that use different systems to manage data. Open data is an initiative to make that data not only available but easily accessible. The findings from the project revealed that the federal government was favouring multinational firms over Canadian small and medium businesses, and even when it did purchase Canadian IT, it chose companies in Ottawa and Toronto. The ThinkData memo recommended that more needed to be done to encourage buying Canadian first and equally across the country.

The reality is that the majority of the Internet consists of non-indexed "Deep Web" that can't be accessed with standard browsers. From a data perspective, it is the proverbial iceberg. The Deep Web shouldn't be confused with the Dark Web where anonymity of users is the first priority. But rather, the Deep Web contains impressive amounts of data residing behind firewalls, paywalls, and dynamic databases that are inaccessible by search engines such as Google. To put size into context, it is estimated that 96 percent of the Internet is Deep Web.[8] One of the most exciting and subtle implications of 5G is the activation of the Deep Web.

Also significant is how education will be revolutionized. Large-scale videoconferencing with participants from around the world should challenge the existing designs and systems for education. It is possible that government education systems could make the leap into an interactive educational environment that will bring the best teachers to the masses without physical restrictions, to a world where it will be feasible to participate in meetings virtually using wireless headsets.

From a corporate perspective, collaboration should improve considerably. Imagine hardware designers in Canada being able to show their latest renderings to manufacturers in China using virtual reality — the pinnacle of intellectual property free trade but with security controls in place. Designers would be able to build individual parts in isolation, then demonstrate how they connect to create products, while manufacturers could subsequently provide feedback in real time and control models, as well.

VR applications consume huge amounts of data and are truly immersive, though at the moment they're quite limited without a wired connection. The networks and devices that utilize 5G will permit VR and new methods of commercial collaboration on a colossal scale. And given time, components will become smaller and more affordable so that a headset will be akin to a pair of glasses rather than bulky Frankenstein electronics.

The longest-term driver of real-time demand for any network is augmented reality or AR. One of the interesting implications of Apple's A11 Bionic chip is that when combined with a network like 5G it will be capable of interacting with the entire world. Augmented reality is exactly what it sounds like — the addition of information and knowledge into a real-world environment.

LATENCY IMPROVEMENTS THAT REDEFINE THE SOFTWARE EXPERIENCE

One area often underappreciated in past network upgrades has been latency. For 4G the latency was approximately 100 milliseconds, while the objective for 5G is to reduce it to under one millisecond. As a point of reference, a reaction to a sudden unforeseen event such as a pop-up ad on a computer can take one second before the user closes the window. Reacting to sound, such as a phone call, can take 100 milliseconds. A reaction to an event that is being paid attention to, but where the outcome is uncertain, such as watching a movie, can take 10 milliseconds.

The 99 percent reduction in network latency to one millisecond will result in a new generation of application potential. If the typical human-computer response to a phone call is 100 milliseconds, then by the time the user has accepted the call, the device and network will have communicated many times back and forth. Before the call is even answered, the app will have already directed multiple queries to social networks and will have determined that the incoming call is perhaps from a relative likely travelling and in need of funds.

The subtle implications of M2M communication in application development is what makes latency easy to underestimate. The fundamental design of software will have to be reconceived as well as the entire user interface on devices in order to allow machines to anticipate future needs. This, in turn, will require the cloud to become more than just a central storage of resources so that M2M interaction can be attained.

For a practical example, consider autonomous driving. The vehicle communicates with the network, which then links with a central repository of information, i.e., the cloud, which connects the vehicle sensor data in real time with a centralized intelligence. All of this is already being done now.

Now imagine a 5G environment where autonomous driving becomes more reliable and effective as cars and infrastructure communicate with each other. Cities are already moving toward modernizing roads and signage, adding intelligence to infrastructure. Soon stop signs will detect when cars are at an intersection and relay oncoming traffic information on demand. Roads will be able to collect data on weather sensors embedded in the pavement, and soon cars will communicate with these devices to retrieve information on highway conditions. In addition, vehicles will be able to talk to one another to better gauge the environment extending beyond their view.

A recent partnership between the map application Waze and the City of Toronto highlights a lightweight approach already in use. On a smartphone, Waze directs how to get to a destination. The unique difference between Waze and other mapping tools is that it's community-oriented, allowing users to leave comments about road conditions. In this partnership, Waze provides the city with traffic data gathered from its half-million customers, while the city supplies Waze with data on road and lane closures, offering better directions for users. Such partnerships are only the beginning of what private-to-public smart cities will accomplish. Upgrades in existing infrastructure to make old devices "smart" will then deliver more real-time information on accidents, weather conditions, et cetera, that can then be fed to mapping applications such as Waze to benefit institutions and citizens alike.

Municipalities will have an important role to play in the 5G future because they will ultimately decide how quickly road infrastructure will be upgraded to meet these needs. If they can accept the vital part that M2M communication will have in application experience, then they'll be open to its logical consequences. All user interfaces will have to be redesigned, including crosswalks and road signs. If the machines are going to have one to a hundred communication exchanges about what we do before we actually do it, then this clearly changes the way options are presented and services rendered. which is why all software has to be revamped to accommodate 5G.

MASSIVE INTERNET OF THINGS (IOT)

Two network specifications of 5G are especially unique: (a) a capability for massive density of computing objects such as those found in urban areas; and (b) a 90 percent reduction in power consumption that allows remote field operation with long-lasting battery life for IoT devices for as many as 10 years. In fact, many of today's popular innovations are part of IoT:

- smart home solutions (intelligent thermostats, locks, blinds, appliances)
- wearables (watches, bands, headsets, shoes)
- smart vehicles (autonomous vehicles, assisted driving technology)
- robotics (smart factories, robotic surgery)

It's important to note that IoT can also include non-intelligent or "dumb" objects that are now "smart." For example, adding sensors to a chair to calculate sitting time and transferring that information over the Internet into a health-and-wellness application is considered IoT. Another instance is a tree with more than 4,000 Twitter followers providing daily data on its breathing and atmospheric conditions.

As a result, IoT has become a booming market, and Ericsson Research is consistent with other market leaders when it believes there will be more IoT devices than mobile phones in 2018.[9] By 2021, Ericsson forecasts there will be 16 billion IoT devices out of 28 billion connected devices.[10]

As mentioned earlier, with billions of new connected devices projected to come online, 5G will be implemented to accommodate non-critical, low-powered ones found in buildings, transportation fleets, and consumer applications (smart homes). These will be in the millions, will be tightly packed together, and will have to last for years before being replaced.

A smaller portion of these devices will be mission-critical and will necessitate ultra-reliability and availability. These include autonomous cars, remote robotics, and traffic safety devices, all of which must have constant connections.

To accommodate both non-critical and critical categories, 5G will likely have to prioritize devices based on how essential they are to public services and enterprise systems.

GOVERNMENT POLICY AND 5G

The genius of 5G is its ability to move across network frequencies to find underused areas and then intelligently spread a connection across an array of spectrum. Most spectrum of economic value and usefulness has been driven by cellular bands, but that represents a small fraction of what's available. This fact opens up possibilities in less valuable areas, such as television or radio, to be converted to the highest and best use.

Governments are not only the regulators of safety policy but they, and the many bureaucratic entities that serve them, are also typically the owners of spectrum, a key resource in 5G. Most people are aware that governments around the world have raised hundreds of billions of dollars from the auctioning of spectrum, mostly to mobile carriers. Spectrum is to mobile what uranium is to nuclear power. Organizations purchase rights to transmit signals over bands of the electromagnetic spectrum.

As a result, governments are major players in the execution of 5G. They can dictate how much spectrum is given out, what frequency bands will be allocated, and how auctions will be conduct-ed. Innovation, Science and Economic Development Canada has already started this process through public consultation that began in June 2017 and will continue until 2020.[11] The focus will be on high-frequency bands in the range of 28 gigahertz, 37–40 gigahertz, and 64–71 gigahertz. These bands can provide the gigabyte network speeds required for enhanced mobile broadband and ultra-low laten-cy. But the downside is that higher frequency bands can't travel far and aren't adequate enough to enable IoT, especially in rural areas. Subsequently, the Canadian government initiated negotiations in August 2017 for the 600 megahertz band, which is a low-frequency one that can stretch into rural regions.[12] This band is more ideal for IoT devices, particularly those that don't require high-data through-put and may need to be accessible in hard-to-reach places.

Governments are the gatekeepers with the keys to resources that will drive the mobile economy, which gives them a strong influence on how the market evolves. Currently, in Canada, mobile markets are controlled by the triopoly of Rogers Communications, BCE Inc. (formerly Bell Canada Enterprises), and Telus Corporation. The federal government believes there isn't enough competition in the market and has proposed setting aside 43 percent of high fre-quency for wireless providers with less than 10 percent subscriber market share. Along with this, there will be a rule that spectrum can't be flipped for five years the way Québecor's Vidéotron re-cently did when it sold unused spectrum from the last auction to Rogers and Shaw Communications.

To put into perspective how high the stakes are in this pro-cess, the U.S. Federal Communications Commission (FCC) set aside 30 megahertz of spectrum for players that didn't have sig-nificant low bandwidth already (i.e., AT&T and Verizon), then conducted a 600 megahertz auction in 2017.[13] T-Mobile was the highest bidder in the United States, paying almost US$8 billion for 1,525 licences across the country. The auction brought in a total of close to US$20 billion. The Canadian government has scheduled its own 600 megahertz auction for March 2019 and expects to rake in at least CAD$1.5 billion with opening bids but is likely to earn much more before closing.

Governments have found utility and applications for the various frequencies of spectrum. Long waves with low amplitude such as radio have been around for a long time, with all stations hosting at a specific tuning point. Satellites operate in a very different area of the electromagnetic spectrum, and governments have attempted to best fit the application with the right spectrum. The reality is that spectrum assignments have been politically influenced. Even more, as our understanding of wireless has evolved, changes in spectrum policy have resulted in a terribly confusing arrangement of resources.

With 5G, network and device intelligence improves vastly, giving it the wide variety of spectrum it needs. The result is not a small increase in efficiency but rather an enormous enhancement that will put tailwinds into all downstream applications.

HOW LONG UNTIL 5G?

New communication standards have long development cycles. Technology needs to be developed, specifications must be agreed on, operators have to do trials within existing networks, and deployment often starts on a small scale. As mentioned previously, since the first automated cellular network in 1980 (first generation or 1G), a new network is launched every eight to 10 years.

- 1979: 1G
- 1991: 2G (11 years)
- 2001: 3G (10 years)
- 2009: 4G (8 years)
- 2020: 5G (10 years?)

The reality is that 5G might be a bit closer than we think. Many operators have begun trials for it, including AT&T, Verizon, Rogers, Bell, and Telus.[14,15,16] We are likely to see larger-scale pre-commercial tests in 2018, while large operators such as Verizon claim they will make initial commercial deployments at the end of 2018.

FIFTH-GENERATION CHAIN REACTION

The 5G network represents a significant evolution in speed, latency, and ability to achieve massive scale with the addition of networked sensors. Many of the revolutionary technologies dominating the media today, including AI and IoT, will be dependent on 5G to achieve their true potential and scale. The result will be a proliferation of devices tied to the Internet.

The relevancy of an increase in connected devices relates to a theorem developed about the value of networks. Metcalfe's law states that the value of a telecommunication network is the square of the number of connected users.[17] Two computers joined to each other are one network connection. Double that to four computers and the network linkages jump from one to six.

Approximately five years ago, the Internet hit 12 billion network connections and it was clear that the number of additions was slowing everything down. This growth was predominately driven by new smartphones. With the advent of IoT and then 5G, we will see an even more dramatic increase in network connections, with most experts predicting a total of 50 billion within the next few years.

As the number of users expands, the value of a network increases exponentially. With 5G we are likely to witness a leap in data consumption never seen previously. Billions of new IoT connections combined with existing scalable technologies such as cloud computing will create an unprecedented explosion in data, unleashing ever more innovation across every industry, all building on the core attributes of 5G and resulting in a chain reaction that delivers exciting new software user experiences for the entire world.

Generation	Primary Services	Key Enablement
1G	Analogue phone calls	Mobility
2G	Digital phone calls and messaging	Secure, mass adoption
2.5G	Digital phone calls and messaging	Increased capacity
3G	Phone calls, messaging, data services	Initial smartphones, first mobile application
3.5G	Phone calls, messaging, broadband data services	Powerful smartphones and app stores
4G/4G Long-Term Evolution (LTE)	All IP services	Rich application that can handle video streaming

* * *

ESSENTIAL TAKEAWAYS

Ray Sharma and Amir Bashir delve deeply into some of the biggest technological advancements that will hit the world in the next two or three years: the end of latency; the introduction of networks so quick they'll enable true AI development; the end of regional isolation in an increasingly urbanized planet; and the delivery of new real-time services in health care, education, transportation, and so much more. Here are some key takeaways from this chapter:

1. *Government investment in technology has now emerged as a basic human right for citizens.* In today's digital society, governments must now continuously invest in technology to serve their citizens. These investments were formerly for economic development purposes; now they're a basic human right. Governments not investing in things such as 5G, artificial intelligence, and data are simply not serving their citizens' needs.

2. *A government must design services for the future, not for current technology.* Too often we hear about how "legacy systems" are ruling how governments design solutions or services for citizens. In today's reality, governments must design services and programs with their citizens and industry in lock step because the world around us is changing too rapidly. My personal opinion on this topic is that we must fundamentally change how governments engage with stakeholders for this to take shape.

3. *The public sector must get used to operating in environments it knows nothing about.* We now live in an age when things will be done to governments whether they're ready for them or not. This is the age of big data, artificial intelligence, borderless states, and mature interconnected societies. Later chapters address how governments are reacting to these realities, but in many cases, we continue to apply analogue thinking to digital realities. We must evolve our thinking to imagine digital first or risk becoming obsolete to the people we serve.

NOTES

1. Jon Brodkin, "AT&T Trialling 5G, Promises Speeds 10 to 100 Times Faster Than LTE," *Ars Technica*, February 15, 2016, https://arstechnica.com/information-technology/2016/02/att-trialling-5g-promises-speeds-10-to-100-times-faster-than-lte.

2. "Canada's Mobile Economy Is Maturing," *eMarketer*, August 31, 2016, www.emarketer.com/Article/Canadas-Mobile-Economy-Maturing/1014417.

3. Edoardo Maggio, "The Chip Inside the iPhone X and iPhone 8 Makes Them More Powerful Than a 2017 Macbook Pro," *Business Insider UK*, http://uk.businessinsider.com/a11-bionic-iphone-x-more-powerful-than-a-2017-macbook-pro-2017-9.

4. Lee Rainie and Andrew Perrin, "10 Facts About Smartphones as the iPhone Turns 10," *Pew Research Center*, June 28, 2017, www.pewresearch.org/fact-tank/2017/06/28/10-facts-about-smartphones.

5. "Cisco Visual Networking Index: Global Mobile Data Traffic Forecast Update, 2016–2021 White Paper," Cisco, March 28, 2017, www.cisco.com/c/en/us/solutions/collateral/service-provider/visual-networking-index-vni/mobile-white-paper-c11-520862.html.

6. "Visual Networking Index IP Traffic Chart," Cisco, accessed April 7, 2018, www.cisco.com/cdc_content_elements/networking_solutions/service_provider/visual_networking_ip_traffic_chart.html.

7. "Complete Visual Networking Index (VNI) Forecast," Cisco, accessed April 7, 2018, www.cisco.com/c/en/us/solutions/service-provider/visual-networking-index-vni/index.html.

8. "How Big Is the Deep Web? A Complete Guide About the Deep Web," *Deep Web*, December 23, 2016, www.deepweb-sites.com/how-big-is-the-deep-web.

9. "Ericsson Mobility Report," *Ericsson*, June 2016, www. ericsson.com/assets/local/mobility-report/documents/2016/ Ericsson-mobility-report-june-2016.pdf.

10. "Ericsson Mobility Report."

11. "Public Consultation on 5G for Faster Mobile Networks," Government of Canada, June 5, 2017, www.canada.ca/en/ innovation-science-economic-development/news/2017/06/ public_consultationon5gforfastermobilenetworks.html.

12. Emily Jackson, "Ottawa Proposes Setting Aside 40% of Wireless Spectrum for New Entrants in Next Auction," *Financial Post*, August 4, 2017, http://business.financialpost. com/telecom/ottawa-proposes-setting-aside-40-of-wireless-spectrum-for-new-entrants-in-next-auction.

13. Jon Brodkin, "T-Mobile Dominates Spectrum Auction, Will Boost LTE Network Across US," *Ars Technica*, April 13, 2017, https://arstechnica.com/information-technology /2017/04/t-mobile-dominates-spectrum-auction-will-boost-lte-network-across-us.

14. Rose Behar, "Rogers, Bell and Telus Are Ramping Up Testing 5G Network Technology," *MobileSyrup*, May 24, 2016, https://mobilesyrup.com/2016/05/24/rogers-bell-and-telus-are-ramping-up-testing-on-5g-network-technology.

15. "AT&T Expanding Fixed Wireless 5G Trials to Additional Markets," AT&T, Aug 30, 2017, http://about.att.com/story/ att_expanding_fixed_wireless_5g_trials_to_additional_ markets.html.

16. "Successful 5G Pilot Places Canada at the Forefront of Global Wireless Innovation," *Marketwired*, June 23, 2017, www. marketwired.com/press-release/successful-5g-pilot -places-canada-at-the-forefront-of-global-wireless-innovation -tsx-t-2223519.htm.

17. Margaret Rouse, "Metcalfe's Law," TechTarget, accessed April 7, 2018, http://searchnetworking.techtarget.com/ definition/Metcalfes-Law.

REFERENCES

"The Tactile Internet," ITU, August 2014, www.itu.int/dms_ pub/itu-t/oth/23/01/T23010000230001PDFE.pdf.

4

Building the
Future of Learning

John Baker

JOHN BAKER is a serial entrepreneur from Waterloo, Ontario, and is the founder and CEO of Desire 2 Learn (D2L), one of the world's leading digital learning firms. His enthusiasm to change how we learn and develop is prominent in everything he does. He has grown D2L from an idea to a large multinational institution with clients in every corner of the globe. John discusses the future of learning, the need for continuous development, the drivers for new types of digital learning, and essentially how government plays a direct workforce management role at both operational and strategic levels. He has changed the learning environment for the better as far back as the days when professors didn't use computers to teach. For him it has been a continuous improvement journey, and readers will see this experience reflected in how he highlights the importance of redefining learning in a digital age.

— Alex Benay

I've spent nearly my entire adult life working and thinking about making learning experiences better — more precisely, about how to create better learning experiences that can serve as a foundation for solving the world's problems. I think learning is the foundation upon which all progress and achievement rests. No matter what problem we are trying to solve — education, politics, the environment — these challenges have learning at the core of building sustainable solutions for them.

Frankly, this single-minded focus on learning isn't what I thought I'd dedicate my professional life to. When I was mapping out my career aspirations, my primary interests were engineering and technology. In the fall of 1995, I enrolled at the University of Waterloo to study systems design and set out to become a skilled engineer with the intention of being a doctor who could also leverage or build technology to save lives.

Although engineering is about problem-solving, I discovered that a better use of my time as a student was identifying the right problems to solve. That point was hammered home for me when my insightful third-year systems design professor gave us an unexpected assignment. He asked us to define a problem — any problem — and then come up with a way to solve it. This was new to all of us; we expected instructors to present the challenges to us, not to ask us to identify problems on our own.

But in the real world problems rarely present themselves as neatly and precisely packaged as a school assignment. They are often messy and ill-defined, and this was our professor's way of encouraging us to be researchers and not just consumers of learning experiences. His open-ended challenge knocked us out of our comfort zone as engineers and forced us to look at our world differently.

It took our team a whole month to land on the one problem we could agree on. Deciding to reinvent crutches, we built and tested prototypes and won prizes for our work.

But that challenge unlocked something in me personally. I never stopped thinking about a bigger challenge. Outside class time, I started asking myself, "What's the most important problem I can solve that will have the biggest impact on the world?"

At the time, my brother was in a distance-education course that required him to listen to cassette tapes of lectures and fax his assignments in. So I asked, "Why can't the Web be used to deliver education to the world instead?"

Then it struck me: delivering a better education experience to the planet could make a huge difference.

Access to education changes lives — it opens opportunities. And I believe no matter what problems we face as humanity, learning is foundational to the answer.

Environmental problems will only be solved through learning. Political struggles will only be resolved through education. Science will only advance through learning. Learning is the foundation upon which all progress and achievement rests. If I could make learning more accessible and more engaging, and in turn, more people could achieve more things because of it, that could change everything!

So, to start, I knocked on the doors of some professors and asked if they would trust me to put their courses online. I got five to agree.

That began a professional and personal journey to address what I think is a critical question: how can we use technology to reach every learner, and in doing so, transform the way the world learns?

IN THE BEGINNING

I started a company to pursue this idea — D2L — that has since grown from a coffee-fuelled start-up to a global company with 750 employees. When we began, just putting courses online was revolutionary. How revolutionary? When we showed professors what was possible, most didn't even have digital projectors in their rooms yet. I'd have to print my presentation slides on acetate overheads so that I could present what a course on the Internet would look like via an overhead projector!

At the outset, we took existing courses and developed methods for instructors to easily make them available online. Universities already offered the kind of distance-learning courses my brother had been taking, and moving these online was a revelation. Course content could be updated quickly. Instructors and students could communicate via instant messaging and discussion boards. Students could communicate with one another, reducing some of the isolation that comes from completing courses at a distance. We didn't just replicate the existing classroom experience — putting it online actually made the learning experience noticeably better. And that was just the beginning.

Technology continues to evolve, and what I've found is that it doesn't merely allow the replication of an existing experience digitally. It also opens up new opportunities as well as new ways of teaching and learning that are more effective. We can now offer educators an array of ever more effective tools that allow them to do more and make learning experiences better.

That's important. Because the world is changing and educators need better tools.

The fact is, if you started teaching 18 years ago when D2L was just launching on its web-based platform, you know that the classroom wasn't the same place it is today.

We are experiencing the biggest transformation of learning in living memory. The classroom of the future will be more personal, engaging, and available, and will equip you with the skills you need for the rest of your life.

And education won't just be the sole domain of traditional schools but will extend into the workplace where learning never stops!

For companies, the learning and development of their employees is rapidly becoming a competitive driver, critical for attracting talent as well as retaining and upskilling existing employees. These shifts point to a world where having access to an effective education is something we'll all need and expect throughout our lives.

DRIVERS OF CHANGE

There are several technological, social, and economic forces driving these changes. One is the increasing pace of technological transformation and its effect on the workforce. The convergence of disruptive technologies, including nanotechnology, artificial intelligence, robotics, genetics, and 3D printing, are reshaping nearly every industry. Dubbed the "fourth industrial revolution," these changes aren't just widespread; they are also happening incredibly fast.

The speed of this revolution accelerates the decay of skills. A generation ago, most people spent the first two decades of their lives acquiring skills that they then applied and refined in the workforce throughout their careers. The fourth industrial revolution makes skills obsolete within the span of a career — arguably several times — while generating demand for new skills previously unimagined. The impacts are daunting:

- Sixty-five percent of children entering kindergarten will end up in jobs that currently don't exist.[1]
- By 2020, more than a third of the desired core skill sets needed by employers will be comprised of skills that aren't yet considered crucial to that job today.[2]
- One estimate, quoted by the World Economic Forum, indicates that nearly 50 percent of the subject knowledge attained in the first year of a four-year technical degree will be outdated by the time the individual graduates.[3]

We will all need to become lifelong learners to keep up, and it's not as if we are currently generating in sufficient numbers the kind of skilled workers needed today. The Georgetown University

Public Policy Institute reported in 2014 that, by 2020, 65 percent of all jobs will require training after high school or a post-secondary degree.[4] And while Canada leads the world, with 56 percent[5] of adults possessing a post-secondary degree, it is still falling short of the attainment levels needed. In the United States, only 45.3 percent of the labour force possesses post-secondary credentials.[6] There is a widening gap between needed and available skills, and unless something changes, it will only get worse.

The nature of work is also changing. A generation or two ago it was common for workers to find "good jobs" and stay perhaps for their entire careers. Today workers are more inclined to shift from company to company or pursue flexible freelance or short-term work arrangements through the so-called gig economy.

The gig economy is bigger than you might think. Already a third of the U.S. workforce is comprised of freelance workers. Some predictions have estimated this number will increase to 50 percent by 2020.[7] During the past decade, gig economy occupations represented 9.4 million individuals, yet over the same period, the U.S. economy only grew by 9.1 million new jobs.[8] If this trend continues, many of us will essentially become our own employers.

The flexibility associated with the gig economy comes at a price. Workers within the gig economy framework tend to have less social protection in the form of rights, are less trained, often face weaker career advancement, and are more insecure about their financial positions.[9] One of the core questions will be how gig workers will find the time and resources to manage their own professional development to ensure they can sustain success in this new economy.

While some workers will find themselves self-employed, the traditional company isn't going away. Yet traditional companies are finding it harder to attract and retain talent.

In the first quarter of 2015, millennials — those born between roughly 1981 and 1997 — became the largest part of the workforce.[10] Millennials tend to be stereotyped as digitally focused, desiring a flexible work schedule, and operating under different incentives compared to previous generations.[11] More important, from a training perspective, they don't stay in the same jobs as long as their parents and grandparents did. While it often takes three to five years to bring a professional to full productivity, many millennials will never hit that threshold.[12] Fifty-eight percent admit they expect to leave their jobs after three years or less.[13] As the need for new skills increases and employees become harder to retain, companies now face a daunting training task.

EDUCATION AND TRAINING: THE NEED TO EVOLVE

Current models are simply not prepared to meet these rapidly changing — and expanding — needs for education and re-skilling. In fact, we are already beginning to see employment opportunities go wanting for lack of candidates with the right skills. More than a third of international companies have struggled to find suitable people for their organizations.[14] Skills such as critical thinking and emotional intelligence are more readily sought after than specific technical skills.[15] At the same time the International Labour Organization estimates that around the world only a quarter of workers have a stable occupation.[16] To the detriment of both workers and employers, we aren't matching skills and needs.

Post-secondary students understand the growing need for skilled workers. That's why many of them pursue higher education. In the 2012 Cooperative Institutional Research Program Freshman survey, 87.9 percent of college freshmen stated that getting a good job was an integral part of pursuing a university degree, approximately 17 percent higher than in the 2006 cohort.[17]

But there appears to be a disconnect. McKinsey & Company note that 72 percent of higher education institutions believe they prepare their students well for the workforce, while half of students aren't sure if their credentials improve their opportunities at finding a job.[18] In another study, only 11 percent of business leaders "strongly agree" that students have the vital skills for the labour market compared to 96 percent of chief academic officers who believe their institutions either are very effective or somewhat effective at providing the necessary skills to students.[19] A survey done by the Association of American Colleges and Universities indicates that 59 percent of students feel they have the necessary skills to succeed in the workforce versus 23 percent of employers.[20]

At the same time the demographics of post-secondary students are changing. As workers realize they need to update their skills, the number of non-traditional students — students older than 25 and often working while going to school — is growing. According to the National Center for Education Statistics, 42 percent of all college students will be over the age of 25 by 2020.[21] These students need a more flexible approach that allows them to focus their limited time on acquiring the new skills they need, not reviewing knowledge they already have.

As needed skills change, we will all become lifelong learners. That not only means we need more flexible post-secondary programs; we also need better learning and development (L&D) opportunities within our jobs. More effective L&D programs help companies deal with millennial-led voluntary turnover. It also helps them keep their employees' skills relevant to their businesses. And, as a bonus, their employees are far more likely to be more productive and engaged.

According to a Deloitte study entitled "Becoming Irresistible: A New Model for Employee Engagement 2015," organizations with a vibrant learning culture had 30 to 50 percent higher employee engagement and retention rates contrasted with other companies.[22] That matters to their bottom lines. It has been reported that roughly US$500 billion is lost in productivity in the United States through a lack of employee engagement[23] — and a

Gallup study found that 68 percent of the workforce didn't feel engaged with their jobs.[24]

Today's education and training model was created in part to prepare workers for an industrialized world where workers repeat defined tasks and follow set routines. As with work in the industrial world, education is defined by seat time, by how long we expect a student to take to learn the content. Once students complete tasks, they move on to the next progressively harder ones, just as we once expected to start our careers with one company and progressively get promoted to greater responsibility. Unfortunately, this is an approach to education that prepares us for a world of work that's disappearing before our very eyes.

THE FUTURE OF EDUCATION

We've reached a point where people will need to access education throughout their lives — at school and through career learning and development opportunities — to maintain in-demand skills. That's a lot more people to educate! We'll need approaches that are scalable, flexible, and efficient to meet this growing need.

The good news is: the same rapid technological advances that are helping to create the demand for continual education and training can also enable new ways of teaching and learning that are capable of meeting growing needs.

Technology gives instructors the ability to reach more learners effectively. Its tools allow instructors to complete common tasks more quickly so they can focus their attention on activities that truly drive learner success. And technology can gather data on how students are progressing that can highlight challenges, show which content generates the most engagement, and help create a more personalized and adaptive learning experience.

A great example of this in action comes from one of my clients, Dr. Jaclyn Broadbent, who teaches a first-year health behaviour class at Deakin University in Australia. Dr. Broadbent has 2,100 students in her class, so she needs tools to help her make sure no student is left out or behind. Compounding the challenge is the diversity of her students. While most are in their first year at Deakin, they can come from as many as 43 different degree programs and have a wide range of academic skills, interests, and pre-existing knowledge.

Reaching every one of those 2,100 students is a daunting task. But with the right tools, Dr. Broadbent has demonstrated that it's possible.

She teaches using my company's Brightspace platform, employing our Intelligent Agents tool, which utilizes event triggers to assist teachers in reaching out with personalized messages. For example, if students don't log into the system for a few weeks, they receive e-mails from Dr. Broadbent. Each e-mail is personalized and the whole system is automated so she doesn't have to spend time monitoring students or manually sending e-mails.

Struggling students who perhaps scored 50 percent on their first assignments and 60 percent on their second ones, who would now be at high risk of dropping out, could receive e-mails congratulating them on the 10 percent improvement. These e-mails would also provide insights into resources that could aid them in their next assignments, while encouraging the right behaviours by telling them how they can seek help. Plus, they get the sticker of the week — often a picture of a cat doing something funny to keep the tone light and encouraging.

Dr. Broadbent also uses our tools to provide timely feedback. Despite the size of their class, students receive feedback within two weeks of submitting assignments and at least a week before their next assignments are due. Feedback is provided via video, so they hear directly from Dr. Broadbent. And there is no need to be perfect on-camera — a cough during the recording is considered "authentic." Not having to record the perfect piece of feedback makes leaving it with audio or video a potentially faster way to give students insights compared to writing comments on paper.

In the end, a tool is most effective in the hands of a master, and Dr. Broadbent is a great teacher. But using technology allows her to engage with students at a scale that wouldn't otherwise be practical. And she's having an impact. Despite the size of her class, students score feedback in her class at 4.72/5, compared to the university average of 3.9. Plus, more than 90 percent of students complete her course compared to a typical 60 to 70 percent completion rate in other large classes.[25]

Dr. Broadbent is an impressive example of the kinds of things we can do today to scale and personalize education. But the great thing about technology is that it's always improving. When it comes to transforming education, we're just getting started.

As we look into the future, machine learning and artificial intelligence — ironically, the very technologies driving disruption in the workforce — will make these tools more powerful and allow instructors to focus less on administration and more on teaching. More important, these tools will allow truly differentiated instruction at scale, offering students content and learning pathways particularly suited for their interests and capabilities.

These tools also facilitate new ways of teaching and learning. Today's education model is still very much based on the concept of seat time. A course is defined by a series of content and tasks that take, say, a semester. No matter how much (or little) you know when you start the course, no matter how much time you need to master new concepts, everyone gets the same amount of time. Technology enables a better approach.

We can begin the shift from a time-based to a competency- or outcomes-based model of learning in which students are quickly evaluated to determine what they already know and what they need to know and don't move on until they demonstrate mastery of new skills, however long that takes. The focus is on not demonstrating how long you learned, but what you actually know, a much more accurate measure of learning.

Without technology tools, evaluating students as they progress is a daunting task and certainly one that would be challenging and expensive to accomplish at any significant scale. With the tools we have today, and technologies that are emerging, a competency-based approach isn't just possible; it's arguably better.

Students can progress through programs, focusing only on the knowledge they need. That means they can potentially move through programs quickly, and in the process, save money. That's what a study of the Texas Affordable Baccalaureate (TAB) Program, a competency-based bachelor of applied arts and sciences (BAAS), found. Compared to traditional studies, TAB students can complete a bachelor degree more quickly, with a savings in tuition and fees of up to 53 percent.[26]

Saving time doesn't just save money. It also creates opportunities. Technology can enable students to learn settled knowledge — the core content and concepts that need to be understood and mastered by a learner — rapidly (sometimes twice as fast)[27] and with better proficiency. They can use that now freed study time to master other skills such as entrepreneurship, sports, music, research, or other areas where they can strive to help us tackle the challenges that lie ahead and develop their full potential.

Technology can also transform a traditional classroom by integrating tools in a blended learning approach, which allows instructors to focus time in the classroom on activities such as discussion or hands-on group activities that are best accomplished in person. Content consumption can happen through access to engaging course materials. It's an approach that leverages the administrative effectiveness and personalization that technology can provide, carving out time for personal and group interaction that assists in consolidating learning. The end result is better learning outcomes. According to a meta-analysis of existing studies by researchers at

SRI International, purely online learning has been found to deliver equivalent effectiveness to face-to-face instruction, and blended approaches have been even more effective than instruction offered entirely in face-to-face mode.[28]

Another transformative approach is to make education available to people for whom it is currently out of reach. Today we're seeing promising progress via massive open online courses or MOOCs, which leverage the growing pervasiveness of the Internet, particularly via mobile, to bring quality education to new audiences.

A great example of the potential impact today can be seen in a MOOC in our platform that allows educators in Tasmania to reach about 25,000 students each semester from around the world. What's more, about 72 percent of the students who start courses on this MOOC, finish them. As technology improves and more people gain affordable access to the Internet, the possibilities of providing entry to education to far more people increases.

The technologies that provide increasing flexibility and personalization in schools will also help companies keep their employees' skills sharp through more effective learning and development activities.

PUBLIC POLICY CAN DRIVE CHANGE

Technology is already helping education meet evolving needs and become more accessible, engaging, and effective. But effective public policy can help accelerate this change.

The Social Sciences and Humanities Research Council in Canada has asked a critical question as one of the six priority areas of research to imagine Canada's future: "What new ways of learning, particularly in higher education, will Canadians need to thrive in an evolving society and labour market?"[29] According to an Institute for Public Policy Research paper, "An Avalanche Is Coming: Higher Education and the Revolution Ahead," there are three major public policy challenges in the education sphere:

ensuring education leads to workforce participation, unshackling the bonds between cost and quality of education, and changing the learning system to support the future of work.[30]

To meet these challenges, we need co-operation between the stakeholders who will drive learning throughout our lives — education institutions, teachers, governments, and companies.

It starts early. We need to work together to make sure the education our kids are getting builds the foundation required for a lifetime of learning. That means making the necessary investments to give teachers the technology tools they require to reach every learner: broadband infrastructure as well as the hardware and software tools necessary to make education engaging and personalized. Investments in professional development are vital to enable teachers to use these tools effectively.

These investments will support a blended approach that brings together the best of both worlds: the flexibility of online, digital tools, and the connections created between students and teachers in classrooms. It's an approach that's perfect for the kind of project- and inquiry-based learning that gives students hands-on experience in asking smart questions, then conducting independent research to find answers — foundational skills for a lifetime of learning to come.

Post-secondary education is no longer simply a next step after high school. As people seek to continually update their skills, it will be a place many of us return to throughout our lifetimes. Colleges and universities need assistance to adapt to this new role and provide the kind of flexibility we will all require.

One way to do this is to craft incentives that support post-secondary institutions that increase access to education and training via online programs, which can reach more learners with high-quality, flexible courses. Two other areas where public policy can drive education innovation are regulations and accreditation. Current regulations favour the conventional university model,[31] which, as I outlined previously, is still very much rooted in the traditional "seat-time" model. Creating incentives to move away from a seat-time approach and toward learner-centric programs such as competency-based education (CBE) will allow non-traditional learners to focus on learning what they need to know and not what they already know.

We also need to help post-secondary institutions make lasting links with industry that create expanded opportunities for experiential learning such as co-operative education programs that mix in-class learning with opportunities to apply learning in industry as part of a student's educational experience. It's an approach that McKinsey & Company note can help lessen the cost of training or education borne by everybody (instructors, students, *and* employers), so there is a lot of value to be found in working together.[32]

While many of us will return to formal education institutions as lifelong learners, we'll still need to learn on the job. It means the line between traditional education policy and skills development policy may blur, and it will likely make sense to create increased incentives for companies to offer enhanced experiential and work-based learning. These are a win-win situation in which workers get to continually develop professionally as part of their existing employment, while companies can hone the skills of their employees to meet emerging market challenges and improve the engagement of their teams.

Governments also play an important, direct role in workforce development programs. They can and should be an example in this shift toward a more learner-centric approach. Moving their training and skills development programs online would allow for greater access, flexibility, and scalability.

EDUCATION MATTERS

While statistics continually show the transformative power of education, most of us have also experienced that power personally. So many of us have had a teacher who inspired us or a moment where we found a subject we were so engaged in that it started a lifetime of personal or professional pursuit.

Not only have I experienced it myself, I've also seen these moments in others. Both of my parents and one of my grandfathers were teachers — the kind students rushed up to thank 20 years after graduation. They worked tirelessly to reach students, help them dream bigger, and assist them to find those transformative moments of engagement and discovery.

All three taught in a small school in a small town on Canada's east coast. Many of their students went on to university; some became doctors, engineers, scientists, and teachers. Their former students still visit them today or invite them back to reunions 20-plus years later. The tiny school punched way above its weight because it had teachers who cared deeply about each student.

I dream of extending this kind of experience to every learner — wherever they are, whatever stage of life they are in. The only way to do that is to support brilliant teachers and give them the powerful tools to reach *every* learner.

* * *

ESSENTIAL TAKEAWAYS

John Baker's chapter highlights key points concerning education in the digital age. The world is moving at a pace in which education should probably no longer be seen as "education" but rather constant re-skilling. In the public service, we too often send people on week-long courses in which everyone sits in a classroom and "learns." That's not to say this model is wrong. But can exclusive classroom learning really keep up with the fourth industrial revolution taking hold around the globe? Here are some of the key takeaways from this chapter.

1. *Governments are educating people for jobs and skills that will most likely not exist.* The pace of technological change is so drastic that the curricula, whether they be government employee training or elementary school systems, is simply not able to keep up with the disruption. We need a fundamental rethink of how we educate and train our citizens in this new reality.

2. *Governments the world over must adapt to the "gig economy."* The very nature of work has changed. Governments can no longer recruit on the promise of a 25- to 35-year career and job stability. And newer generations of employees aren't necessarily looking for stability but rather job satisfaction. What this means for a slower, more linear educational and training model isn't clear yet. One thing that is certain is that a constant rethink of the education and training landscape is now essential to how public services support, deliver, and take training, courses, and skills development workshops.

3. *Fresh public policy is needed.* This strikes at the heart of one of the main issues discussed throughout this entire book. How are governments supposed to lead through traditional "policy" processes when the pace of change makes government policies outdated from the moment they're written? Governments need new thinking and policies in the education space to remain relevant as nations. The countries that empower their citizens the fastest with some of these new realities will be able to take globally dominant positions regardless of size.

NOTES

1. "The Future of Jobs: Employment, Skills and Workforce Strategy for the Fourth Industrial Revolution," World Economic Forum, January 2016, www3.weforum.org/docs/WEF_Future_of_Jobs.pdf.
2. "The Future of Jobs," 20.
3. "The Future of Jobs," 20.
4. Anthony P. Carnevale, Nicole Smith, and Jeff Strohl, "Recovery: Job Growth and Education Requirements Through 2020 — Executive Summary," Georgetown Public Policy Institute, 2014, https://cew.georgetown.edu/wp-content/uploads/2014/11/Recovery2020.ES_.Web_.pdf.
5. "Adult Education Level," OECD, accessed April 7, 2018, https://data.oecd.org/eduatt/adult-education-level.htm#indicator-chart.
6. "A Stronger Nation," Lumina Foundation, April 2016, www.luminafoundation.org/files/publications/stronger_nation/2016/A_Stronger_Nation-2016-National.pdf.
7. Brian Rashid, "The Rise of the Freelancer Economy," *Forbes*, January 26, 2016, www.forbes.com/sites/brian-rashid/2016/01/26/the-rise-of-the-freelancer-economy.
8. "The Future of Jobs," 17.
9. "Policy Brief on the Future of Work: Automation and Independent Work in a Digital Economy," OECD, May 2016, www.oecd.org/employment/Policy%20brief%20-%20Automation%20and%20Independent%20Work%20in%20a%20Digital%20Economy.pdf.
10. Richard Fry, "Millennials Surpass Gen Xers as the Largest Generation in U.S. Labor Force," Pew Research Center, May 11, 2015, www.pewresearch.org/fact-tank/2015/05/11/millennials-surpass-gen-xers-as-the-largest-generation-in-u-s-labor-force.

11. Krista Jones, "What Millennials Really Want from Employers," *Financial Post*, December 3, 2015, http://business. financialpost.com/executive/careers/what-millennials-really-want-from-employers.

12. Jon Bersin, "Spending on Corporate Training Soars: Employee Capabilities Now a Priority," *Forbes*, February 4, 2014, www.forbes.com/sites/joshbersin/2014/02/04/the-recovery-arrives-corporate-training-spend-skyrockets.

13. "New Study: The 2015 Millennial Majority Workforce," *Upwork* (blog), October 29, 2014, www.upwork.com/ blog/2014/10/new-study-2015-millennial-majority-workforce.

14. "Unleashing Greatness: Nine Plays to Spark Innovation in Education," World Economic Forum, July 2016, www3. weforum.org/docs/WEF_WP_GAC_Education_ Unleashing_Greatness.pdf.

15. "The Future of Jobs," 22.

16. Graham Lowe and Frank Graves, *Redesigning Work: A Blueprint for Canada's Future Well-Being and Prosperity* (Toronto: University of Toronto Press, 2016), 19.

17. Michelle Weise, "Got Skills? Why Online Competency-Based Education Is the Disruptive Innovation for Higher Education," *EDUCAUSE Review*, November 10, 2014, http://er.educause. edu/articles/2014/11/got-skills-why-online-competencybased-education-is-the-disruptive-innovation-for-higher-education.

18. Dominic Barton, Diana Farrell, and Mona Mourshed, "Education to Employment: Designing a System That Works," McKinsey & Company, January 2013, www.mckinsey.com/ industries/social-sector/our-insights/education-to-employment-designing-a-system-that-works.

19. Weise, "Got Skills?"

20. Hart Research Associates, "Falling Short? College Learning and Career Success," AACU, January 20, 2015, www.aacu.org/sites/ default/files/files/LEAP/2015employerstudentsurvey.pdf.

21. Weise, "Got Skills?"

22. Josh Bersin, "Becoming Irresistible: A New Model for Employee Engagement," *Deloitte Insights*, January, 26, 2015, https://dupress.deloitte.com/dup-us-en/deloitte-review/issue-16/employee-engagement-strategies.html.

23. "State of the American Workplace: Employee Engagement Insights for U.S. Business Leaders," Gallup, 2013, www.gallup.com/services/176708/state-american-workplace.aspx.

24. Amy Adkins, "Majority of U.S. Employees Not Engaged Despite Gains in 2014," Gallup, January 28, 2015, http://news.gallup.com/poll/181289/majority-employees-not-engaged-despite-gains-2014.aspx.

25. "Deakin University: Engaging Every Learner," D2L, 2018, www.d2l.com/success-stories/deakin-university.

26. Carlos Rivers and Judith Sebesta, "Competency-Based Education: Saving Students Time and Money," *EDUCAUSE Review*, December 12, 2016, https://er.educause.edu/articles/2016/12/competency-based-education-saving-students-time-and-money.

27. Marsha Lovett, Oded Meyer, and Candace Thille, "The Open Learning Initiative: Measuring the Effectiveness of the OLI Statistics Course in Accelerating Student Learning," Carnegie Mellon University, 2012, https://oli.cmu.edu/wp-content/uploads/2012/05/Lovett_2008_Statistics_Accelerated_Learning_Study.pdf.

28. Barbara Means, Yukie Toyama, Robert Murphy, and Marianne Baki, "The Effectiveness of Online and Blended Learning: A Meta-Analysis of the Empirical Literature," *Teachers College Record* 115 (March 2013): 35, www.sri.com/sites/default/files/publications/effectiveness_of_online_and_blended_learning.pdf.

29. "Imagining Canada's Future," Social Sciences and Humanities Research Council, accessed April 7, 2018, www.sshrc-crsh.gc.ca/society-societe/community-communite/Imagining_Canadas_Future-Imaginer_l_avenir_du_Canada-eng.aspx.

30. Michael Barber, Katelyn Donnelly, and Saad Rizvi, "An Avalanche Is Coming: Higher Education and the Revolution Ahead," Institute for Public Policy Research, March 2013, http://med.stanford.edu/smili/support/FINAL%20 Avalanche%20Paper%20110313%20(2).pdf.

31. Barber, Donnelly, and Rizvi, 20.

32. Barton, Farrell, and Mourshed, "Education to Employment," 21.

5

On Openness

Hillary Hartley

HILLARY HARTLEY, another change agent, is one of the co-founders of 18F in Washington, D.C., a team designed to "do" government differently, with the goal of delivering user-centric digital services. Her entire career has been dedicated to digital disruption. Recently, she moved to Canada to become chief digital officer for the Government of Ontario, where her change-management skills will be put to the test to modernize the provincial government's operations. On a personal front, I look forward to working with Hillary and her team over the next few years as both levels of government continue to create stronger ties. Here, Hillary focuses on the need to work in the open if governments want to be serious about becoming "digital." What this means is that public servants around the world must get used to developing policy, solutions, and services in the open where everyone can see them. This new concept is necessary for serious digital democracies, and Hillary explains why. Gone are the days of analogue policy and program development.

— Alex Benay

G overnments won't be spared the exponential change we're seeing in other facets of our modern lives. As we shift toward an economy of abundance, it's imperative that governments around the world take note of the ways they can recognize their own areas of abundance. The next wave of making governments simpler, faster, and better will be one that follows this trend and understands how to make it scale.

Just as ride-sharing apps such as Uber or Lyft leverage the abundance of available drivers and cars, or Airbnb a wealth of available real estate, governments should start recognizing what assets they have to offer to the sharing economy. Whether it's unused or underused real estate, parts of the workforce that may have extra cycles due to automation or digitization, or the vast amount of data they collect, governments are clearly poised to benefit from positioning themselves as "exponential organizations."

However, just as with the recent buzzwords "digital transformation," the real barrier to this shift isn't rules or regulations. It's also not a lack of desire on the part of the public sector or even defined policy outcomes. It's the ingrained and entrenched culture that often surrounds information and service delivery, which enables well-meaning public servants to keep doing things the way they've always done them.

Change management is hard, and *Titanic*-sized ships turn slowly. If we're going to make governments simpler, faster, and better, we have to create a culture in which leaders understand their roles not only in managing specific programs and projects but also in modelling the types of behaviours that help create effective and efficient organizations in this age of abundance. Government leaders must become digital leaders — folks who understand that it's not just defining different outcomes that's necessary but also making changes to their organizations' structures, practices, values, and behaviours. As Janet Hughes, former program director for GOV.UK Verify at the United Kingdom's Government Digital Service (GDS), explains:

To make this kind of transformation happen you need people at the top who understand that it's not just about doing better digital projects, adopting the latest new technology, having a Twitter account, or even automating your existing business processes, or bringing in some technical experts. It's about completely changing the way you think and work as a leader, and the way your whole organization works (including, in some cases, considering whether the organization needs to exist at all in its current form).[1]

Also, in one of the best blog posts of 2017 for the digital government community, Janet Hughes wrote, "Digital organizations are responsive, open and efficient."[2] A digital organization (or one that can operate effectively in the Internet era) is able to understand and respond to rapidly changing needs or expectations. If done well — meaning, there are clear goals, free-flowing information, and an abundance of trust — being digital should reduce cost and risk and make your organization more efficient. But to be fully responsive and bear the fruit of efficiency, your whole organization needs to work in an open way.

There are several things that are key to creating an environment in which a digital organization can thrive, including

- a focus on user needs at the very heart of the organization;
- an agile mindset allowing you to iterate based on those needs;
- being prepared to fail and understanding that means you're learning;
- empowered, enabled, and equipped teams who are trusted to deliver;
- evidence-based and data-informed decision-making; and
- a willingness to ask questions and challenge assumptions.

Each of these principles can help those "people at the top" navigate the culture change that must go hand in hand with digital transformation. And while a focus on user needs absolutely must lie at the heart of your organization, the linchpin of it all is openness. To paraphrase self-described open government and data ninja Pia Andrews: open without digital doesn't scale, and digital without open doesn't last.

LEADERSHIP

Digital leaders — and, as I mentioned earlier, we're all digital leaders nowadays — need to understand that the values and behaviours of their organizations are as important as the information and services they provide. The people at the top of the hierarchy need to model the behaviours that will enable their teams to do the hard and important work of being open, people-centred, and adaptable in the face of change.

Sameer Vasta, digital anthropologist and my colleague at the Ontario Digital Service (ODS), distills these values for digital leadership in the public service into eight principles:

1. **Obsess about the user.** Make every decision with the end-user, the person who will use your product or service, in mind. Demand that of your team.
2. **Be agile and iterate.** Get feedback on and make changes at every step of the process. Change and update based on ongoing feedback and user research.
3. **Work out loud.** Encourage your team to communicate the process to co-workers, to clients, and to partners throughout the work, not just when there is something to announce.
4. **Use the data.** Rely on measurable evidence to inform your decision-making and hold your team to the same standard. Make sure all decisions have mechanisms for measurement.

5. **Be prepared to fail.** Failure is inherent to risk-taking — as long as the failure is small, iterative, and drives to a better user experience.

6. **Challenge everything.** Ask questions and don't assume things need to be done in a certain way because they always have been. Many implicit "rules" are actually customs; let your team challenge those customs.

7. **Embrace the chaos.** Your team members will have different optimal methods of work when it comes to time, place, environment, tools, and process. Encourage them to work in their best ways, as long as the work and team dynamic aren't compromised.

8. **Be unreasonably aspirational.** Because the work matters and the public deserves it.[3]

Transformation demands leadership — in both position and action. It demands that, as leaders, we grow comfortable outside our comfort zones, and requires that we all take responsibility for making changes to our organizations' structures, culture, and practices. There's a thread of openness throughout this list, whether that means working to flatten power structures, embrace trust, and empower teams to make decisions; rewarding experimentation, collaboration, and working out loud; or modelling servant leadership in which leaders exist to support, coach, mentor, and clear blockers for their staffs.

Servant leadership has come to be rather closely identified with agile product management in which, ideally, there's no concept of hierarchy, and instead, leadership revolves around removing barriers and empowering and enabling teams to get their work done. With a focus on service as opposed to hierarchy, organizations can be more responsive to the needs of their people and more efficient in the delivery of products and programs.

However, sometimes we fail — and, if we're truly being agile, lean, and iterating, hopefully we're "failing small" with each development sprint — which is when a servant leader can shine. There are some leaders who will get very upset and try to ferret out blame; there are other leaders who will act as if nothing happened and not address the mistake at all. Neither reaction helps move an organization forward, and it can cause massive performance or morale hits to the team.

The servant leader sees failure from a wholly different lens: failure — big or small — is fundamentally a learning opportunity. Treating failure as a "teachable moment" helps create a work environment where we actually learn from our mistakes. When we also imbue this environment with a spirit of openness and transparency, we're able to allow not just our own but an entire ecosystem of teams to *only make new mistakes*. Unless we are sharing our failures along with our successes (whether it's simply internally across teams or publicly for the good of the ecosystem), projects will continue to be delivered in the same way and with the same flaws. When we're open, everyone can benefit.

MAKE THINGS OPEN — IT MAKES THINGS BETTER

There is an argument to be made that I wouldn't be writing this chapter, or that digital service teams around the world — teams such as 18F, U.S. Digital Service, Australia's Digital Transformation Agency, Ontario Digital Service, to name some of the direct diaspora — wouldn't exist if not for the openness, vociferous blogging, and rigorous documentation culture of the United Kingdom's GDS. The British went first, and not only did they pave the way, they left us well-documented instructions, templates, and lessons learned.

Working out loud enables teams to document their successes and failures, to transcribe their history in real time, and to allow other teams to learn from those "teachable moments." GDS has enabled countless other government teams not to start from

scratch because of the work it did documenting its plans, processes, and principles. In fact, its official government design principles
conclude with "Make things open, it makes things better":

> We should share what we're doing whenever we
> can. With colleagues, with users, with the world.
> Share code, share designs, share ideas, share inten-
> tions, share failures. The more eyes there are on a
> service the better it gets — howlers are spotted,
> better alternatives are pointed out, the bar is raised.
>
> Much of what we're doing is only possible be-
> cause of open-source code and the generosity of the
> web design community. We should pay that back.[4]

OPENNESS AND DESIGNING WITH, NOT FOR

"Open" and "government" have gone hand in hand for years now:
open government, open source, open data, open standards, open
engagement, open dialogue, open information, and so on. Each
of these developments has pushed all levels of government toward
greater transparency, participation, accountability, and accessibility.

Since the first small city states appeared 5,000 years ago, people have needed information to hold their governors accountable
and make informed decisions. The evolution of democratic governments, the idea of free and fair elections, combined with freedom
of expression and freedom of the press, have enhanced this need
for timely information. When all the people of a state are involved
in the political and policy decision-making process, information
is key. And now "open government" and an "open-by-default"
mentality in government and policy at all levels of government are
pushing this forward into new domains.

Open government teams are looking across their jurisdictions to remove organizational barriers and enable better

decision-making. When government data, information, and dialogue are open by default, we enable a more transparent, accountable, and accessible public service to emerge.

Openness can provide a mechanism for trust. We shouldn't mistake openness as a proxy for trust, but we can use it as a tool to help rebuild public trust and re-engage people with their governments. Open government and participatory design strategies are a path forward that can address both governance issues and delivery.

For too long too much of government operations, policy-making, and delivery have happened behind closed doors. In recent years, we've seen a movement in governments across the world from an inward-looking, "government knows best" approach to a user-centred view that designs and delivers public services from the perspective of the end-user. Governments have taken tools they've used for policy development — the consultation, the town hall, reaching out to constituents for advice — and have begun using those same ideas for service development and information delivery.

This user-centred view of policy and service delivery is integral to transformation, but it's also not enough. We must also embrace true collaboration with people and develop policy and services *with* them, not just for them.

A truly human-centred view of government requires that we, quite literally, meet people where they are. We first began learning this with government's adoption of social media nearly 10 years ago. Those years were full of baby steps, a toddler beginning to understand a new type of direct engagement — one that wasn't necessarily driven by government but that was often initiated from the public. We learned to "fish where the fish are," taking our messages out of direct mail and advertising to Twitter and Facebook to have open, honest, and often hard conversations with people wanting to understand government.

Getting outside our bubbles is a necessary component to the consultation strategies of governments, and the same rubric applies to understanding the users of our services and the consumers of our information. But if we're truly building with, not for — if we're co-designing services with the people who will use them — then we have to go beyond the typical consultation models and beyond researching users. Co-design recognizes that people who need information or services are the subject matter experts of their lived experience. And that while user research is important, those subject matter experts (our users) can and should help us actually design the service or online experience.

Recent research on service design suggests that teams create more innovative concepts when collaborating with customers and end-users than they do when working on their own.[5] When design teams give people the tools to be creative, share insights, and envision their own ideas, they are able to combine their "expert" knowledge with lived experience. The conversion of these two sets of knowledge can quickly generate better ideas with higher perceived user value; improve the team's knowledge and understanding of the user need; get quick validation of ideas or concepts (or, alternatively, "fail fast"); lower development costs and reduce development time; and encourage co-operation among people, organizations, disciplines, et cetera.[6]

PUBLIC-SERVICE REFORM CENTRED ON OPENNESS

The public service has come to increasingly rely on the problem-solving potential of the crowd. In Ontario in 2017, Secretary of the Cabinet Steve Orsini published his vision, "Transforming the Ontario Public Service for the Future." Central to that vision is a willingness to create new approaches that focus on co-design and collaboration with the public, using open source and open data to supercharge those efforts. Ontario has framed its open government initiative around three pillars of work:

1. Giving Ontarians more opportunities to weigh in on government decision-making
2. Sharing government data online so everyone can help solve problems that affect Ontarians every day
3. Providing Ontarians with the information they want and need to better understand how their government works

The Open Government Engagement Team expands on these three elements — open dialogue (public engagement), open data, and open information (transparency) — based on a report it prepared by consulting with Ontarians:

- **Open Dialogue** is about using new ways to provide the public with a meaningful voice in planning and decision-making so government can better understand the public interest, capture novel ideas and partner on the development of policies, programs, and services.
- **Open Information** is about proactively releasing information about the operation of government to improve transparency and accountability and promote more informed and productive public debate.
- **Open Data** is about proactively publishing some of the data collected by government in free, accessible, and machine-readable formats and encouraging its use by the public as well as within government.[7]

The team concludes that "Generally the more closed any aspect of a system is, the greater the extent to which responsibility for that aspect falls upon a single entity, and thus is introduced as a single point of control."[8]

Again, openness is a mechanism for building trust. Rebuilding trust in government is a crucial component of the success of these efforts. And it's not just government.

Many of the services in the platform-and-sharing economy suffer from a trust issue. Facebook's fake news problem could have a real impact on democracy. Twitter is used as a platform for extremism fuelled by bots and fake accounts. Uber settled a U.S. Federal Trade Commission investigation into data mishandling, privacy, and security complaints by agreeing to 20 years of independent privacy audits. After years of racial discrimination claims on Airbnb, the service now allows the California housing regulator to test and penalize Airbnb hosts for racial bias.

There is a fundamental issue of trust that these companies have to decide how to address. What values will guide that decision-making? Will it be a laser-like focus on the user? Working out loud? Opening up their data and application programming interfaces (APIs)?

The sharing economy must, in a sense, go back to its roots. Trust has always been what keeps our economy growing. Moreover, trust was the foundation that let companies such as Uber, Airbnb, and TaskRabbit exist in the first place. In fact, they've designed for it, giving us reviews, ratings, and other cues that explicitly signal various proxies for trust. As stated in PwC's Consumer Intelligence Series report *The Sharing Economy*, "It's the elixir that enables us to feel reassured about staying in a stranger's home or hitching a ride from someone we've never met."[9]

It's conceivable that the companies that will come out on top are the ones that convince us they're the best stewards of our information and data — the platforms clearly designed with inclusion top of mind. We've gotten comfortable in today's "perk economy" where we hand over our data without hardly a second thought for some semblance of convenience. But the day isn't too far away when our collective desire to understand and trust these companies will outweigh convenience. The only way companies will be able to execute on that, to really convince us, will be through a renewed spirit of corporate openness — through open dialogue, open data, or open information.

Trust at scale is hard. But I believe governments are uniquely situated to show the path forward. They're designed for resilience. They're designed for consultation and decision-making. And with the injection of openness and transparency that we're seeing in public sectors across the globe, governments can be early models of what the trust economy at scale really looks like when it delivers digital services.

OPEN SOURCE/OPEN BY DEFAULT

In an effort to default to open, the U.S. Digital Service Playbook states:

> When we collaborate in the open and publish our data publicly, we can improve Government together. By building services more openly and publishing open data, we simplify the public's access to government services and information, allow the public to contribute easily, and enable reuse by entrepreneurs, nonprofits, other agencies, and the public.[10]

The U.S. Digital Service Playbook, developed by the United States Digital Service and 18F in 2014, contains 13 "plays" that define the path to good digital government service delivery. Number 13 is "default to open," and it's an important note to end on because it is the linchpin to improving government together.

Publishing data in the open and through open application program interfaces both simplifies access to government services and information and allows people and groups outside government to serve audiences and communities that might be hard to reach. And, as stated on Data.gov, the U.S. open data portal:

Open government data is important because the more accessible, discoverable, and usable data is, the more impact it can have. These impacts include, but are not limited to: cost savings, efficiency, fuel for business, improved civic services, informed policy, performance planning, research and scientific discoveries, transparency and accountability, and increased public participation in the democratic dialogue.[11]

The Canadian government's open data portal website identifies the following benefits of open data:

Support for innovation — Access to knowledge resources in the form of data supports innovation in the private sector by reducing duplication and promoting reuse of existing resources. The availability of data in machine-readable form allows for creative mash-ups that can be used to analyze markets, predict trends and requirements, and direct businesses in their strategic investment decisions.

Advancing the government's accountability and democratic reform — Increased access to government data and information provides the public with greater insight into government activities, service delivery, and use of tax dollars.

Leveraging public sector information to develop consumer and commercial products — Open and unrestricted access to scientific data for public interest purposes, particularly statistical, scientific, geographical, and environmental information, maximizes its use and value, and the reuse of existing data in commercial applications improves time-to-market for businesses.

Better use of existing investment in broadband and community information infrastructure — Canada has invested in information and communications networks in the form of technical infrastructure and community services, such as libraries and social service agencies. This investment will continue to add value-for-money for Canadians by extending Web technology from a one-way communications medium to a collaborative environment.

Support for research — Access to federal research data supports evidence-based primary research in Canadian and international academic, public sector, and industry-based research communities. Access to collections of data, reports, publications, and artifacts held in federal institutions allows for the use of these collections by researchers.

Support informed decisions for consumers — Providing access to public sector service information to support informed decision-making; for example, real-time air travel statistics can help travellers to choose an airline and understand the factors that can lead to flight delays. Giving Canadians their say in decisions that affect them and the resulting potential for innovation and value (builds trust and credibility).

Proactive Disclosure — Proactively providing data that is relevant to Canadians reduces the amount of access to information requests, e-mail campaigns, and media inquiries. This greatly reduces the administrative cost and burden associated with responding to such inquiries.[12]

When prioritizing digital services, one of the questions I wrestle with is *What should governments build or deliver versus what should they enable?* Many public servants understand the potential power of open government data, and indeed, entire industries have been built on it — global positioning systems, weather, to name two. But when deciding how to digitize a new service, it's worth considering starting with an API. Creating an open API allows governments to make data available, control access and privacy, and enable the creation of services both internally and externally.

18F highlights that "Providing information and services through Web APIs supports interoperability and openness. Well-designed APIs make data freely available for use within agencies, between agencies, in the private sector, or by citizens."[13] It lists the following as benefits of using APIs:

- **Increase the reach of your services** by allowing other agencies, partners, and the private sector to integrate — and amplify — your agency's data and content.
- **Save time** through automation. You can update data or content once, and your API can refresh in multiple locations automatically on a website, mobile platforms, and on social media venues.
- **Save costs** by allowing third-party innovators to use information and services to create new, useful products that are beyond the scope — or budget — of your agency.
- **Speed product development** through improved prototyping and ease of access for internal teams and sister agencies by allowing granular and open access to content.
- **Build markets** by improving access to government resources such as health, economic, energy, education, and environmental data for entrepreneurs to build upon.[14]

Underlying all this work must be a foundation of openness, and open source and working in the open are the bedrock of that foundation.

There's a misconception that open source is just a type of software licence that gives access to free code. However, the "free" in free open-source software (FOSS) refers to freedom, not cost. The licence is simply the mechanism by which the creator allows the source code to be used, updated, or shared under certain terms and conditions. For individuals and organizations involved with open-source projects, the community and the broader set of values are often more important than the particular licence. Open-source teams, like many government digital teams, embed the philosophies of openness and collaboration into their projects from idea to launch.

Opensource.com explains that to do things "'the open source way' means expressing a willingness to share, collaborating with others in ways that are transparent (so that others can watch and join too), embracing failure as a means of improving, and expecting — even encouraging — everyone else to do the same."[15] Bringing this open-source mentality to government is necessary in order to realize any intentions around open data, open information, or striving to be open by default. Government is full of source code — documentation, policy, templates, et cetera — that guides the way we interact with and understand it. When we make this "source code" (whatever its form) open, accessible, and shared, many people can have a hand in adapting or modifying it for the better.

To sum up everything I've said previously, I'll borrow, once again, the wise words of Pia Andrews:

> The public service holds and creates a lot of data in the process of doing our job. By making data appropriately publicly available there are better opportunities for public scrutiny and engagement

in democracy and with government in a way that is focused on actual policy outcomes, rather than through the narrow aperture of politics or the media. This also builds trust, leads to a better informed public, and gives the public service an opportunity to leverage the skills, knowledge and efforts of the broader community like never before.[16]

* * *

ESSENTIAL TAKEAWAYS

Hillary Hartley has shown us why governments must change their mindsets from analogue, where paper processes trump engagement and often leadership, to institutions where the power of the Internet, collaboration, and the information age are seen as levers for all levels of the public service. The transition of public institutions from a time when computers didn't exist to the current digital age isn't occurring at the same speed as the changes all around us (think artificial intelligence, blockchain, the cloud, the Internet of Things, et cetera). This chapter reveals important factors to consider as governments make the shift to "working in the open."

1. *Leadership is paramount.* From government ministers to individual analysts, we must get used to working in the open to regain public trust. Showing and engaging with the public on what we're working on, whether it's a policy or a program, is critically important because it fosters dialogue.

2. *Developing with, and not for, people is important.* In fact, it may be *the* key. It's about governments regaining their empathy in how we serve. It means co-developing policies, programs, and services out

in the open with others. We must go to where people are and stop expecting people to come to us as
governments. The cellphone and the Internet have
killed this notion of people coming to us, yet government mentality hasn't necessarily adjusted to this.

3. *Open by default is the future of government.* Governments must get accustomed to developing
platforms where others can engage, create, and
develop. Using open-source technologies permits
a "developing-with-not-for" approach. Opening up
data stores permits companies, researchers, or everyday citizens to engage with their governments to
build better democracies, create economic growth,
and ultimately develop better countries.

NOTES

1. Janet Hughes, "Digital Leadership: Changing Your Whole Approach, Not Just Doing Better Digital Projects," *Doteveryone* (blog), Medium, February 10, 2017, https://medium.com/doteveryone/digital-leadership-changing-your-whole-approach-not-just-doing-better-digital-projects-6e75389f938b.

2. Janet Hughes, "What a Digital Organisation Looks Like," *Doteveryone* (blog), Medium, June 6, 2017, https://medium.com/doteveryone/what-a-digital-organisation-looks-like-82426a210ab8.

3. Sameer Vasta, "Unreasonably Aspirational: Leadership in a Digital Age," *Ontario Digital* (blog), Medium, October 24, 2017, https://medium.com/ontariodigital/unreasonably-aspirational-leadership-in-a-digital-age-8dda4d07e0.

4. "Government Design Principles," GOV.UK, April 3, 2012, www.gov.uk/guidance/government-design-principles.

5. Jakob Trischler, Simon J. Pervan, Stephen J. Kelly, and Don R. Scott, "The Value of Codesign: The Effect of Customer Involvement in Service Design Teams," *Journal of Service Research* 21, no. 1 (2017): 75–100.

6. John Chisholm, "What Is Co-Design?" Design for Europe, accessed April 7, 2018, http://designforeurope.eu/what-co-design.

7. Open Government Engagement Team, "Open by Default — A New Way Forward for Ontario," Government of Ontario, December 24, 2015, www.ontario.ca/page/open-default-new-way-forward-ontario.

8. "Open by Default."

9. "The Sharing Economy," PWC, April 2015, www.pwc.com/us/en/industry/entertainment-media/publications/consumer-intelligence-series/assets/pwc-cis-sharing-economy.pdf.

10. "Default to Open," *Digital Services Playbook*, U.S. Digital Service, accessed April 7, 2018, https://playbook.cio.gov/#play13.

11. "Impact," Data.gov, accessed April 7, 2018, www.data.gov/impact.

12. "Open Data 101," Government of Canada, December 19, 2017, http://open.canada.ca/en/open-data-principles.

13. "Introduction to APIs in Government," Developer Program, accessed April 7, 2018, https://api-all-the-x.18f.gov/pages/introduction_to_APIs_in_government.

14. "Introduction to APIs in Government."

15. "What Is Open Source?" Opensource.com, accessed April 7, 2018, https://opensource.com/resources/what-open-source.

16. Pipka, "Creating Open Government (for a Digital Society)," *pipka.org* (blog), September 18, 2012, http://pipka.org/2012/09/18/creating-open-government-for-a-digital-society.

6

Government Social Media

Jennifer Urbanski

BECAUSE THIS BOOK is about all things digital, it is appropriate to mention that I have never met Jennifer Urbanski face to face, only virtually through social media. Jennifer, a dedicated executive at LinkedIn, is based in Toronto, Canada, where she works with federal departments, provincial ministries, crown corporations, economic development organizations, and advocacy groups to drive engagement with Canadians and professionals worldwide. She has worked in the telecommunications industry at Rogers Communications and has been a digital advocate for years. I have worked personally with Jennifer on increasing the use of social media in government, not because it is wonderful to have but because the world has become digital and social media, in many ways, is now its language. Jennifer takes us through concepts such as user experience, how to employ each channel appropriately, and the importance of real two-way dialogue and not utilizing social media as simply a type of content "push" mechanism, among other topics. The good news is that she is easy to find online, so if readers have any questions, they should feel free to seek her out!

— Alex Benay

No one has to convince anyone that our world is digital or that more people are using mobile phones and living in newsfeed environments. We already know this. We've heard it at conferences, we've seen it in research, and we've discussed it in meetings.

There are a few examples that point in the direction of how governments should engage with citizens on social media, but none have achieved the level we need to be at — a level that aligns with the growing expectations our modern society has for digital experience.

As a marketer — someone who has worked on the sales and business development side of things for the past 10 years — I've seen requests from many clients or media agencies looking for a "big idea," something that's *never been done before*. To be honest, though, most of the time, such requests are deemed unnecessary given the plethora of existing solutions, or are seen as difficult to achieve given each party's capabilities and limitations. But what's exciting about working on the effective use of digital and social media by governments is the rare opportunity to actually do something new and unique. And it doesn't have to be a big idea. The basics can be put into practice to make better digital leaders, and *that's* the big idea. It's the changes that are made to improve how we operate that establish departments or ministries as innovative, and for the first time in history, we have the chance to truly be leaders in social government.

THE USER EXPERIENCE

Let's start by looking at what social government really is. I like to think of it as the modern-day town hall where a community comes together to hear its government discuss the issues it cares about most. Governments have always been fond of public consultations, especially town-hall meetings that communicate important information and ask citizens for feedback on key issues. But in-person meetings pose many challenges — location, scheduling, frequency, and willingness to speak out in front of others, to name a few. Social media present a tremendous opportunity for governments to collect feedback from citizens on a variety of issues at scale and to stay connected with their communities. Embrace such opportunities and don't make them complicated. I've seen departments run public consultation campaigns online that require complex registrations and log-in processes. Stop that! Digital statistics have shown over the years that the more steps taken to complete an action, the more likely it will be abandoned.

Now, to be sure, many people will want to pounce on this and argue that to make everything work we need to validate that people are actually citizens of a country or fit within the demographic we're trying to reach. Again, stop! What is more important — a handful of opinions from people who actually bother to go through all the complicated steps to complete a registration process, or thousands of contributions from engaged citizens with a few unqualified submissions thrown in? Stop doing what has always been done and justifying it with excuses. *Look* at the data. As already mentioned, the data proves that a complicated user experience causes a significant decline in user responsiveness, so why continue to make things difficult for users?

The following are characteristics of effective social media activity:

- builds awareness and attention to the work being done by the department or ministry
- distributes important messages
- gathers valuable feedback and perspectives
- unites a community

The problem is that government organizations globally are still living in the past and are creating experiences rooted in traditional tactics. All too often social media posts look like flyers stapled to trees announcing annual town halls rather than rich images with impactful messaging that suit the current newsfeed environment. In addition, the process and procurement systems aren't aligned with modern technology and the pace citizens move at. Governments around the world need to revisit their policies and procedures and strip out inefficiencies if they want to succeed in the modern world.

The great news is that people who want to bring change and are interested in how to make a difference are showing up more and more in governments. The areas to focus on to pursue this change are: hiring the right people, giving them the right tools, and creating a simple process that enables change to occur.

LOOKING AT THE "PEOPLE PART" FIRST

When hiring communications, marketing, or social media staff, make sure they truly know social media and digital. This doesn't just mean they use Facebook every day in their personal lives or tweet five times a day. Insist that they bring examples of successful campaigns they've run to interviews and cover the following:

- How have they grown in the social space and what have they learned?
- What mistakes have they made and what have they learned from them?
- What reputable training have they completed?
- How would they innovate your department's digital presence and what ideas can they bring to the table?

If the hiring manager or recruiting team doesn't have the skill set to test for the above, then form a hiring committee with people from within or outside the department who have proven they're knowledgeable and successful in this area. Hire staff who showcase passion, which, admittedly, can be tough. Finding employees who are passionate and truly care about advocating for their company on a daily basis can be like winning a lottery. It's what we all hope for but few actually find. It's essential to do this, however, because people who are passionate about an organization are the voices behind a brand.

TOOLS THAT EMPOWER STAFF

Since the voices of employees are such important factors in the success of overall objectives, make it easy for staff to use those voices. Understand the department's social media policy and ensure that staff are clear about the details. If such measures aren't taken, people might not feel empowered to participate in social conversations for fear of breaking the rules.

The best social media playbook is from the U.K. government, which fully embraces and acknowledges the importance of public servants participating in digital space.[1] It's easy to read, uses aesthetically pleasing formatting, and feels modern. The playbook presents a strong case for using social media in the public sector and covers everything from community management to creative and content advice, and it summarizes all the social media channels in which the U.K. government is active. The tone of the content is empowering and concentrates on the positive ways public servants can succeed digitally. This is in contrast to many government social media policies that tend to focus on rules and instill fear in public servants who are interested in using social media channels. What is particularly inspiring about this playbook is that its guidelines were published in 2014, which shows how much the U.K. government is ahead of the rest of the world. When I read this information, I'm inclined to jump out of a chair and say, "See? This is what we've been telling you all along!" It's frustrating when many people say the same thing, but the message still isn't being listened to and implemented.

This message needs to be heard by anyone interested in improving digital approaches within governments. The most success comes from teams that work together and collaborate effectively. They ask questions when they don't know something, they share information, they make the time to speak to one another, and they take risks together in the pursuit of improvement. This is how to achieve goals and truly innovate. Look past egos and stop thinking that everything is already known. Ninety-five percent of global government teams aren't doing this right. Focus on getting better, listen to the advice being offered, and apply it in the workplace. As stressed earlier in this chapter, employing the basics is the big opportunity right now. That's how differences can be made.

Some government teams believe they've posted effective guidelines for their staffs, but they really haven't. Trying to navigate the Government of Canada's social media guidelines[2] is a challenge just to find the information online, and when it is located, it's very copy-heavy with little formatting and a lot of dry jargon that can be confusing for people who aren't experts in digital media. The same can be said for the social media guidelines published by the U.S. government through its General Services Administration site.[3] First, a PDF has to be downloaded to access information that's already too complex for a smooth user experience. Next, the guide is very policy-focused with the majority of content centred on role definitions, how to minimize risk, all the rules needed to follow, and way too many hyperlinks connecting readers to different sources of information. This resource needs a complete overhaul so that it easily houses all information in one place and encourages a positive and empowering opportunity for public servants to use social channels. Certainly, governments need to include policy and rules, but the United Kingdom has proven that these needs can be acknowledged while still empowering staff to participate in online communities.

Once social media guidelines have been revamped, spend time sitting down with employees to review the details and answer questions. Staff members have to be advocates for the work their department or ministry is doing. Employee networks are often 10 to 20 times bigger than a government department's, so use this resource when trying to engage with the public. If employees ask questions whose answers aren't known, commit to finding solutions and then communicate them to the whole team, not just the individuals who asked.

CASE STUDY: THE ONTARIO CABINET OFFICE

In my experience over the past two-plus years working on government marketing campaigns, one of the inspiring teams blazing the trail is the Ontario Cabinet Office. In 2016 it was tasked with building out a new social media strategy for the Government of Ontario and outlined its success in an article posted on *Medium* by Jennifer Stubbs, senior manager of marketing and digital strategy.

> We've believed, for a long time, that Ontarians should be able to get information in the places where they live online. That belief is at the core of the work the Cabinet Office social media team, a team charged with developing content, guidelines and strategies for the Government of Ontario social channels with a renewed focus on client service. The team also provides advice and support to all ministry social media channels for both paid and earned communication campaigns.[4]

In her article, Stubbs outlines the strategy used to create a team and mobilize support from other staff to drive the change necessary to get Ontario up to a new standard of social communications. Since its inception, her team has grown from one person to more than 80 staff and volunteers.

The first stage of its team-building strategy was to design a focus group of staff from across the Ontario Public Service (OPS) who oversee social media. By listening to their feedback and experiences, a training strategy was designed to deliver a better user experience (UX) for Ontario citizens and set the standard for social media practices across all ministries. Stubbs called this team "The Implementers" because it would inject this new knowledge into social activity moving forward.

The second stage involved a diverse set of volunteers labelled "The Innovators" who were passionate about building a better government. This team didn't have to work in social media — it could hold any position within the OPS. This type of diversity is what any team should strive for, since it ensures different backgrounds, experiences, perspectives, and ideas will be represented in the pursuit of a solution. The Innovators identified areas of innovation and built out the road map to achieve UX goals for Ontario's digital strategy.

But perhaps the most effective piece of the puzzle was harnessing the power of Ontario's senior staff to really push through the great ideas and work being generated by the first two teams. Dubbed "The Advocates," this group of supporters was able to overcome the barriers that had historically existed within government communications. By involving a team of senior staff early in the process, Stubbs and her people were able to keep momentum going on ideas once they reached the highest levels of approval.

This strategic approach of harnessing the right people helped the Ontario government achieve a few major highlights in its first year, including

- building government visibility by standardizing account names and verification on all ministry social media accounts;

- supporting responsible social media use in the OPS by updating the 2013 social media guidelines with criteria for official, professional, and personal use;
- hosting a first-ever social media boot camp for 50 ministry staff; and
- enhancing customer service on social media by launching a research group to review best practices and service standards.

CHOOSING THE RIGHT SOCIAL MEDIA CHANNELS

Once the right people are in place and government staff members feel empowered to advocate for their departments or ministries, the focus then turns to the process or tactical strategy of how to be an effective social government. The first step is to choose the right social media channels, selecting two or three and focusing on doing those right rather than trying to be everywhere at once. The right channel for a message and audience isn't just what is used in personal life. The job of a communications/marketing/digital/ social team member is to be well informed and make choices based on the objectives and needs of the department. A job is best performed by stepping outside of the comfort zone and learning what works for an audience above any personal considerations.

What follows are four key questions to ask when evaluating which channels are right for a department or ministry.

1. Which Channels Target the Desired Audience?

Although pretty obvious, the point here is to choose channels that actually target the desired audience. All too often time is invested promoting content on Instagram or Twitter when the targeted audience is highly unlikely to use those platforms or the message doesn't fit within the context of those channels. Don't jump on a social platform's bandwagon if the expenditure in time can't be justified.

An economic development organization will likely focus on LinkedIn and Twitter as the most effective social channels to reach its intended audience. On the other hand, a health department that wants to speak to mothers about breastfeeding will have a greater chance of success on Facebook.

This process could also involve stretching skills and knowledge, since the right channel might not be used personally by staff members, a reminder that people who know social media beyond personal usage should be hired. One of the biggest challenges I've had in my role at LinkedIn is that a lot of public-sector staff are late adopters of the platform because they remember it in its traditional job-seeking function. However, the evolution of LinkedIn has made it a rich newsfeed environment where influential and aspirational professionals go to have conversations around important topics. Government communications staff who refuse to utilize LinkedIn based on personal bias miss out on the strong engagement seen globally when government content is posted. The obstacle in this case is the person, not the government, who is hindering potential success.

Make sure the target audience is truly understood — this means more than just the numbers or how many are on a certain channel — and make certain a clear persona has been defined and then find out which of the major channels can target as closely to that persona as possible. Also be aware of organic versus paid targeting options, since sometimes there are differences that will affect content strategies.

2. What Is the Right Message to Communicate?

By carefully thinking about the message that needs to be to communicated, the format that will be most effective and the mindset required to resonate with the target audience can be evaluated. In marketing there is a lot of talk about the right message to the right person at the right time, which is exactly the point in asking this question. Think about clothing designers who want to show off their new fall collections. They should choose a social platform heavy in rich, visual content such as Snapchat or Instagram. If those same clothing designers want to talk about the ethical practices of their fabric suppliers, they should consider leveraging LinkedIn and the mindset of those members who are ready to consume information that's a little more serious in nature.

3. What Are the Mandated Gaps or Needs?

Similar to some of the rationale above, identifying the main goals of a department should help navigate channel choices. When I first started working on public-sector business, I gave a presentation — later turned into a long-form post that can be found on my LinkedIn profile[5] — at the Government of Canada's Digital Advertising Day, highlighting the publicly released mandate letters from Prime Minister Justin Trudeau to his cabinet ministers. These letters set the direction for government departments to execute on, and the common goals throughout were open government, transparency, accountability, making data-driven decisions, and professionalism. Those objectives require revisiting social media strategy and considering the right channels for delivering them, which might mean changing what's been done historically to provide a more modern approach. Think about what the unique needs are *before* deciding on the channels to invest time in. Don't try to fit a square peg into a round hole — if a channel doesn't align with the strategy required to achieve the desired needs, don't bother!

A story I once heard made me chuckle. A marketer walked into his colleague's office frustrated because he couldn't figure out how to integrate Pinterest into his social media strategy. When his colleague asked why he was trying to use Pinterest, he replied that he didn't know. He just wanted to use it because he'd heard another department had employed it successfully! Talk about the wrong answer! Only choose channels that fulfill specific needs. Note that this might change at different points throughout the year or for different campaigns and could be different from what other government peers are doing.

4.Which Channels Drive the Most Value?

One of the first things I hear from clients on a daily basis is that the people they report to don't understand the right metrics to focus on in digital marketing. Patiently, I listen to their frustrations trying to explain or justify their digital strategy only to be debated by someone who just doesn't get it. For those in any position of decision-making power or on a team working with data, put aside pride, stop acting like a know-it-all, and make sure digital measurement is understood. Different options will be available and they should ultimately align with the desired objective. Such data might be very new and specific to social and digital media, and there will be slang, abbreviations, and general experience rules to become familiar with. The great news is that there are many people who have gained this experience, are moving into government, and are eager to share this information and make an impact. Ask lots of questions and seek out knowledge to become the most valuable resource to colleagues and country alike.

Most of the time, when seeking awareness for a message, announcement, or service, measurement is calculated by how many people actually click through to a site, not how many impressions are received. However, to ascertain true engagement, say, for a public consultation campaign, measurement of effectiveness will be done by looking at how many comments people made or how many people submitted a comment on the website. That means engagement metrics such as comments or completions are more important than the number of people who saw a website or clicked through to it. Understanding key metrics such as click-through rate, engagement rate, return on investment, share of voice, and lead value are good places to start and should help determine channel choices. Understand what all the possible metrics are and their use cases at a higher level before focusing on which ones specifically indicate success for an initiative. When concentrating on the right metrics and measures properly, the appropriate channels worth time and investment can be quickly determined. Understanding the differences between each channel and the metrics they report on is also important. Digital channels or social media platforms aren't always "apples to apples" and comprehending this stuff is crucial when a team has to make a decision or an update on the team's recent marketing campaign is required.

QUALITY OF EXECUTION

The next step is the quality of execution. If there is one area where government needs to improve, it's in the content being put into social channels.

Creative means everything, and the public is used to the highest quality creative possible given the current state of popular media. Between television and digital, our brains are accustomed to rich colours, clear imagery, concise messaging, and beautiful sounds. Years of marketing research by big private-sector companies have been combined with billions of dollars to ensure the content being shown to the public hits on every psychological queue and physical sense. This is the level that government content needs to be at. It doesn't have to cost a lot of money, either. There are endless solutions to putting high-quality content in front of the right audience, and all it takes is a little effort.

The public-sector leaders who are emerging globally in the creative arena are those in economic development. Higher-quality content is being released across the appropriate channels from governments in the United Kingdom, the

European Union, France, Australia, and the Greater Toronto Area (GTA). They offer the best examples of government teams producing high-quality videos and static images showcasing impactful statistics related to the attractiveness of their regions. They are proving that releasing modern content like this shouldn't go through a complex government approval process and that higher quality results in higher engagement.

An update of branding is likely necessary for most government organizations globally. There are quite a few government logos around the world that are used as avatars or within content that's very old and traditional-looking. One of the key areas of innovation for a social government is to overhaul branding to be in line with modern society. This point fits into some of the others previously made concerning standardization, high-quality creative, and competing with millions of messages consumers see on a daily basis.

Instead of trying to find examples of government campaigns that have strong creative, look at big brands for inspiration. Apple, Coca-Cola, and Nike are all brands that can serve as inspiration to optimize creative. That doesn't mean investing huge production dollars, either. It's merely a matter of sourcing high-quality images and ensuring that content fits platform specifications and suits the context of the chosen social media channel.

Outstanding creative has come from Canadian provincial and U.S. state governments and a handful of both country's federal departments, but even more low-quality creative has been issued from such government teams. It's better to say nothing at all than to present dull, lifeless images to the public. If a strong design team can't develop creative that appeals to people used to bold, bright, beautiful, entertaining, and rich imagery, reconsider spending time and budgets on launching something "just because it's expected." Scroll through Facebook, LinkedIn, Twitter, and Instagram feeds to determine what works, what's weak, what's quickly noticed, and what's too easily breezed past. Often these things can be figured out solely based on personal preferences.

TWO-WAY DIALOGUE

Frequently, government groups like to distribute dialogue that talks *to* users rather than talks *with* them. This happens most often with economic development teams tasked with attracting foreign direct investment or innovation teams trying to speak to citizens about the projects being advanced. All these groups post the same reports shouting that they're the best city or province or country for all the same things. No one is unique and no one tells users *why* they're the best. However, there's a strong community in Ottawa, Canada's capital — in which Alex Benay, one of the co-authors of this book, is an active participant — that's challenging this status quo and creating really effective two-way dialogue with government staff and citizens. A lot can be learned by following and connecting with this community.

Groups who want to stand out and be different should use strong statistics or quotes in their content to communicate why or how they're different from what's been done historically or in comparison to other countries, provinces/states, or cities. Teams should also connect with influencers within their industries and engage in discussions with them. This involves active community management so that responses to comments can be done in a timely manner and engagement with external content can be achieved as soon as it's posted. Content on mass social channels such as Facebook, Twitter, and Instagram disappears into a crowded abyss within 24 hours, which means acting fast is vital in order to stay in the conversation. A channel such as LinkedIn works off a two-week algorithm, so contributions to discussions last a little longer. It's important to understand how algorithms work on each social channel used to know how often to check in and engage.

CORRAL THE COMMUNITY

Using effective hashtags is one of the best ways to get organic social media messaging into the existing conversation taking place around a cause. There's ample existing research on how to do this right. Best practices include keeping hashtags short, simple, and relevant; choosing one or two hashtags and not overusing too many; leveraging existing or trending hashtags to join a conversation or community; and searching hashtags before using them to understand the conversation happening around them. Experts also recommend employing the same hashtag across each social media channel to keep things consistent.

UNDERSTAND MEASUREMENT AND HOW TO MAKE DATA-DRIVEN DECISIONS

Deep data and analytics are usually offered in much more complex partnerships than individuals in governments generally possess in the social media space. The former include media campaigns or partnerships with data providers for use in government tools. Individuals who want to be better at social media will have metrics that are a lot more basic than members of a marketing or communications team have, but personal metrics can still help guide social behaviours.

Individuals will notice patterns in the engagement they get on the things they post. The number of likes and shares, or the quality of comments, helps indicate the interest networks have in the content being posted. Using high-quality images catches much more attention than images automatically pulled from certain landing pages. Adding unique perspectives or thoughts on a topic creates more engagement with contacts than simply sharing links with no colour commentary. As more content is posted and engagement with others is expanded, the basic signals offered by social networks will be used to do more of what works and less of what doesn't.

Those who directly impact the social media success of their specific teams or units will have opportunities to use more sophisticated technology, measurement tools, and data. The challenge posed here is that antiquated communication and privacy policies will need to be revamped, as well.

Understanding and planning measurements in the initial stages of project planning are essential. A strong team will define its goals during the briefing and brainstorming process of the planning cycle. This requires a commitment to those goals and knowing in advance what data to give more weight to once the campaign or project is up and running. It's always amazing how challenging this can be for government teams. One of the first questions we ask our clients at LinkedIn — and other social networks do this, too — is: "What are you looking to accomplish with this campaign?" After that the conversation goes as follows:

- **Client:** "Awareness of issue XYZ."
- **LinkedIn:** "Okay, so you'll be measuring the campaign based on impressions, views, or likes?"
- **Client:** "Well, no, we'll be looking at clicks."
- **LinkedIn:** "Oh, okay, great, so driving engagement and traffic to your site is your main objective?"
- **Client:** "Well, yes, that too."

The strategy behind an awareness versus an engagement campaign will be different. There are different ways to speak to an audience and different words, images, and products to use. A team needs to agree on goals and then work with an agency and/ or vendor to properly structure a campaign to achieve what must be accomplished.

The ultimate tipping point for social government and the ability to effectively evaluate campaigns and gather useful data will be the successful implementation of tracking pixels, which are pieces

of digital code placed on websites to calculate user behaviour. They
track this behaviour at a macro level and don't report on an individual level. The goal of this data is to guide business decisions related
to user experiences, preferences, and accuracy. This unit of measurement allows comprehension, at a high level, of the following:

- Are the right people coming to the site?
- Which social channels are they coming from?
- Where are users spending their time on the site?
- Which information is most important to them?
- Where is the audience being lost?
- How often are users engaged with the content?
- What new opportunities are there to improve products and
  services?

Government teams are terrified of collecting this data properly due to expected backlash from citizens and cyber breaches,
but let's acknowledge the reality that this information is already
being collected and stored somewhere. That's how the Internet
works — it really is a matrix of ones and zeros. Every time actions are completed online, they're being stored somewhere. It's
actually more dangerous that no one is aware of where such data
is collected and stored and how it's really being used, but that's
an entirely different discussion outside social government (and
aligned more with cybersecurity).

Putting systems into place to properly track and analyze data
is essential to being leaders in social government. It would be
useful to share examples of traditional government administra-
tions already doing this, but old policy prevents governments
worldwide from implementing tracking pixels. Revisiting policies to adapt to modern digital practices is a first-mover opportunity for countries that want to be in the vanguard to modernize
government communications.

To check if a government site has tracking pixels on its website, use a free tool such as Ghostery. If the site does, it can immediately be confirmed that the government is focused on modernization in an effort to improve its citizen services. There are crown corporations (similar to government agencies in other parts of the world) that have implemented tracking pixels to guide their business and marketing efforts. In Canada they include Canada Post, Export Development Canada, Canada Mortgage and Housing Corporation, and Destination Canada. All of these organizations properly utilize online data to improve business operations and ensure taxpayer dollars are spent effectively. They take this responsibility seriously and provide best-in-class opportunities for other government teams to follow. To learn more about the work these crown corporations are doing in this regard, reach out to their marketing or legal teams for more details. These teams have done a great job to ensure compliance and protect privacy as they become more innovative and move toward modern ways to service the public.

Economic development organizations will be the next to follow the examples being set by crown corporations, and departments and ministries should follow soon after that. The public shouldn't be scared of this — these tools allow governments to guarantee they are getting the right information in front of the right people when it's needed most. There's no sense sending messaging about a certain topic to people who don't require any information about that topic, and it's a huge waste of tax dollars if money is spent in a way that doesn't provide proof of its effectiveness.

In lieu of these sophisticated tracking methods, government groups turn to basic measurements such as impressions (the number of views on a post), clicks, and reach (the total number of people exposed to a message). The challenge with these metrics is that they lack the quality required to qualify people as the right audience to receive that message. Sophisticated tracking methods involving pixels give deeper insight into ensuring the right people are getting the message and performing the behaviours required to reach intended goals.

Look at the model that Jennifer Stubbs developed with the Government of Ontario to bring change into policy around government pixelating. Establishing a team of strong advocates (senior-level leadership) is essential to achieve this mission.

As governments step into the next phase of digitalization, social media will be a key part of effective communication, citizen engagement, and public consultation. Personally, I can't wait to see the leaders who will emerge in the digital space, because when they do, they will change the course of history forever and establish a new form of democracy.

*   *   *

ESSENTIAL TAKEAWAYS

Jennifer Urbanski promotes the use of social media not as "nice to have" but as one of the most important tools in the shed for governments that want to change how they engage with stakeholders. The days when governments could simply push out content to the masses via websites are over. Even now, governments simply use social media to disseminate content, which is a failure to understand what digital truly is. We now have the tools for true democratic dialogue and too often squander their potential. This chapter guides us through a few cases in which governments around the world have taken leadership positions and have chosen to connect online.

1. *It's okay for government employees to demonstrate passion for their work.* The days of the silent public servant are long gone. In today's digital world, it's expected that governments will engage because

that's the reality of digital citizens around the world, since they can engage with anyone from the convenience of their mobile devices of choice. Therefore, governments must enable their staff to talk and involve themselves with the public more and remove old command-and-control measures when that makes sense.

2. *Successful online engagement requires leadership.* If ministers, deputy ministers, and other senior officials don't engage personally, how can we expect the movement to enable hundreds of thousands of public-sector employees to truly take shape? Perhaps this is a generational issue, or maybe it's merely a time-management problem, since we know how busy ministers and deputy ministers can get. Regardless, today's leaders must engage personally online, otherwise the outreach is a bogus attempt at dialogue. Nowadays, everyone can see through genuine versus fake online "conversations."

NOTES

1. "Social Medial Playbook," *GDS Digital Engagement* (blog), GOV.UK, accessed April 7, 2018, https://gdsengagement. blog.gov.uk/playbook.

2. "Social Media and Web Requirements," Government of Canada, June 15, 2017, www.canada.ca/en/treasury-board-secretariat/topics/government-communications/social-media-web-requirements.html.

3. "Guidance for the Official Use of Social Media," U.S. General Services Administration, accessed April 7, 2018, www.gsa. gov/reference/guidance-for-the-official-use-of-social-media.

4. Jennifer Stubbs, "A Shadow Force for Social Media Modernization," *Ontario Digital*, Medium, April 11, 2014, https://medium.com/ontariodigital/a-shadow-force-for-social-media-modernization-274e4190b525.

5. Jennifer Urbanski, "Trudeau Should Look to LinkedIn if He Wants to Present a Professional Government to Canadians," LinkedIn, November 24, 2015, www.linkedin.com/pulse/trudeau-should-look-linkedin-he-wants-present-jennifer-urbanski.

7

The Future of Digital Government Services

Olivia Neal

I FIRST MET OLIVIA NEAL through mutual professional contacts and had heard about her work bringing standardization of digital practices into play within the government of the United Kingdom. In light of such an accomplishment, I sought her out to join our team in the CIO Office for the Government of Canada. She came as advertised: genuine, caring, smart, and also relentless in upholding her digital principles, all things we needed at the time in Ottawa. Olivia is a University of Leeds graduate and previously held public-sector positions with the Judicial Appointments Commission and the Government Digital Service in the United Kingdom. Here, she guides readers through the complexities of digital service delivery in the future, from multi-platform offerings to the need for culture change, borderless state service strategies, and beyond, clearly demonstrating that there is a lot to do toward designing new-age digital services.

— Alex Benay

As the pace of technology change continues to drive new approaches for the delivery of services, the need for governments to keep up to stay relevant grows in importance. Using examples from around the world, with a particular focus on Canada and the United Kingdom, this chapter explores the evolution of digital government services and how governments can prepare in the future to provide services to their citizens.

WHAT IS A DIGITAL GOVERNMENT?

An organization that believes being digital is solely an issue for its chief digital officer or chief information officer will fail. Being a digital organization isn't about building a better website or even embracing and using the latest technologies; it is about being an organization that knows how to operate effectively in the digital era and how to continually meet the raised expectations of its users.

There are many definitions of "digital" out there. Tom Loosemore, one of the founders of the United Kingdom's Government Digital Service (GDS), offers among the best: "Digital: Applying the culture, practices, processes & technologies of the Internet-era to respond to people's raised expectations."[1]

Loosemore's definition highlights two important areas: (1) digital change isn't something created and done, but rather, it is about building a new way of working to quickly adapt and use new approaches; and (2) the purpose of this is to respond to the raised expectations of people. For governments, this means fundamental changes to traditional ways of operating. The role of governments and the people who work for them is, as it always was, to serve the people of the countries they govern, to protect them through effective defence and justice systems, to regulate and support growth, and to ensure the welfare of citizens through the provision of services.

As with any other organization, the phase of talking about digital government will move on. We will no longer need to say we are digital because it will be inherent in the way we and other organizations operate.

THE EVOLUTION OF DIGITAL GOVERNMENT

By their nature, governments aren't subject to the same pressures as private companies. They are frequently monopoly providers of services that are essential to citizens' well-being or are legally required. As such, the users of government services often have no choice but to use the interfaces provided to them. But that isn't to say governments haven't put into place measures to make delivery more efficient internally, as well as more effective for citizens.

Prior to the introduction of the Internet, options for service delivery were, of course, focused on face-to-face transactions, paper forms to be posted or faxed, and the telephone. In 2002 in the United Kingdom, Jobcentre Plus was created, merging the role of the job centre and benefit provider to give unemployed people access to multiple services in one place and to connect the operations of two linked areas. This built on the role of traditional labour exchanges, which first came into play in the United Kingdom in the early 20th century.

In Canada, Service Canada was created in 2005, bringing together provision of a range of federal services and benefits for individuals such as pensions and employment insurance. Service Canada also became the first point of contact for reaching other federal government services, including a combined call centre for general inquiries. The toll-free number 1-800 O-Canada provides information and a first response, though it still requires people to call a second specialist centre to allow them to complete their transactions.

The International Telecommunications Union (ITU) is the United Nations' specialized agency for information and communication technologies. It publishes statistics on Internet access around the world, which show that prior to 1997 fewer than 11 percent of people in the developed world were Internet users. In 2017, 94 percent of young people (aged 15 to 24) in developed countries were using the Internet, along with 81 percent of the rest of the population.[2]

The statistics also demonstrate that in the developed world, as a whole, active mobile broadband subscriptions have increased from 19 percent in 2007 to 97 percent in 2017. In the same time period in the developing world, this has gone from 1 percent to 48 percent.

Looking back to 2007, in most countries (other than Estonia and a few other outliers) government services were almost entirely paper and in-person-based. Clearly, the rate of Internet adoption since then has vastly eclipsed the rate at which governments have adapted to the digital age.

Some nations, led by Canada in the early 2000s, brought parts of government information together to be available in a single place on the Internet. But there largely remained significant distribution of information across numerous departmental and agency websites, forms, brochures, and PDFs. This meant that for users of government services — for businesses and citizens — services were not only difficult to use but difficult to find, and it was hard to know what to trust.

During the 2000s, the concept of digital-by-default developed, which took on different meanings in different contexts.

In Denmark the government took a bold political decision, and from 2010 mandated government departments to provide, and its six million citizens to use, online services. This became the Danish definition of digital-by-default. The government stopped sending letters in the post and made all government communication to citizens online only. People were able to apply for an exemption from the requirement, but they had to prove need. The result is that over 85 percent of Danish individuals now use government services on the Internet. But that doesn't provide a guarantee that those services meet the expectations of their users.

In 2011, in the midst of the economic crisis, the U.K. government took a different approach. Rather than mandating the use of online services, the mantra of digital-by-default was interpreted as "digital services so good people prefer to use them." Led by the newly created GDS, the government not only developed services with users at their heart but also published the results. All new and redesigned services had to make public information showing user satisfaction, the service completion rate, the cost per transaction, and digital uptake.[3] This more organic approach has generally resulted in a slower uptake of digital services than the mandated Danish model, but it has placed a higher bar on user focus in service delivery.

CHALLENGING THE TYRANNY OF LOW EXPECTATIONS

Governments around the world have started, to a greater or lesser degree, to make individual services better for end-users. The approach of bringing focus to user-centred, redesigned interfaces has often been referred to as "lipstick on a pig." But even if the technology supporting a redesigned service still desperately needs to move on, or the internal processes in a government department still need fixing, if a new service can immediately benefit its end-users, it is worth doing, even if it isn't a truly transformational approach.

By showing how government services can work better for the people who use them, the tyranny of low expectations is challenged. Service redesign, even if it starts with the front end only, has proven that government services can be something not just to be tolerated but to be looked at as a leading example of excellent design.

The way that governments provide services to their citizens doesn't have to be at the absolute cutting edge of what's being offered in the private sector. In most countries it never has been. But it has to be close behind. It has to take advantage of the ability to quickly utilize design approaches and technologies that have been tested in the private sector to leapfrog the initial iteration and adoption phases.

As people's ability to quickly and easily access online services continues to grow, it is increasingly unrealistic to expect them to physically go somewhere and wait to gain access to something to which they're already entitled. This creates anger, disillusionment, and disengagement.

Services that work for the people who use them are crucial for a functioning system. And if governments don't keep up with the rest of the world, then they risk becoming irrelevant.

The opportunity that artificial intelligence (AI) and automation offers for governments to adapt the way they deliver services and run internal operations is large and ever-growing: simultaneous translation into not just the official languages of a country but all languages of users; the ability to automatically tag and curate internal documentation to create easy access and remove the need to rework; and automatic development of new policies — based on evidence, data, and previous work — ready to be tested with users. These are just a few examples. The ability of AI to fundamentally change government operations is transformational. As the introduction of the automated teller machine (ATM) allowed bank staff to focus on higher-value areas of contact, the automation of straightforward tasks will allow government employees to focus on more complex aspects of citizens' lives.

Consolidation of government services is often studied through the lens of the costs it will save. Traditionally, there has been less focus within governments to attempt to quantify failure demand, that is, the costs of services that don't work for their users.

In 2016 in the United Kingdom, the National Audit Office assessed some cost-cutting measures of Her Majesty's Revenue and Customs, designed to make call centres cheaper to run. Where call centre staff were removed before better digital alternatives were put into place, it was found that for every £1 saved by the government, the cost to a user of the services increased by £4, based on a conservative calculation of the value of people's time.[4]

Call centres represent a large proportion of the costs associated with many government operations, and for those running them, the battle is less about call-waiting times and more about reducing the number of calls unanswered. Developing digital services with a relentless focus on the user reduces error rates, especially when data can be pre-populated and manual calculations can be avoided, allowing agents to support those who need it most with complex queries. This trend will be supported by AI and automation, allowing standardized processing and minimal interaction from users.

BUILDING A CULTURE FOR THE DIGITAL ERA

Governments are often the largest employers in a country. And the larger the organization, the harder it can be to drive change throughout it. Within governments, as in all sectors, change is necessary to survive. And it has to be embraced. The purpose of the role of public service is clear. As the pace of technology change develops, governments need to be engaged and leading. It shouldn't just ensure that legislation keeps pace to allow innovation and economic development to thrive; it should also make sure that it takes advantage of those developments for itself.

Rewarding Flexibility and Embracing Change

The culture of governments needs to adapt to reward flexibility and embrace change. The Government Communications Headquarters (GCHQ) in the United Kingdom is an organization in which staying ahead of technology is not optional but vital to national security. It's responsible for working with law enforcement and intelligence agencies to defend the United Kingdom, and to protect the government from cybercrime. GCHQ and its predecessors have been in place for more than a century, and there is little room for error in the work of its 6,000 staff members.

People with the skills GCHQ needs to recruit are in high demand. The organization has recognized the importance of culture to attract the absolute brightest and best to work with it. GCHQ exemplifies an organization that's actively open and engaging while being both relevant and authentic.

GCHQ's approach to openness includes the way it works. One of the most significant things, which allowed GDS to promote the benefits of making code open and treating it as a public asset, was the backing and leadership of the National Cyber Security Centre in demonstrating the value of this by doing it.

Often in governments when there are options available, there is a tendency to overemphasize the risks of change without properly examining the cost of failing to do so. When the National Cyber Security Centre developed its own information technology (IT) system, it blogged about the process it followed. The centre wanted its approach to be an exemplar for others, both within governments and outside. It said "most importantly — the system had to be a pleasure for people to use. We were serious about that being the most important characteristic. A highly secure solution that no-one uses isn't secure at all."[5]

In this approach, the responsibility and emphasis in development were placed firmly on the team leading the development to get it right for the user, and not the other way around.

Building Multidisciplinary Teams

Governments need to work differently in order to build organizations that can continue to embrace new technologies and approaches, as well as quickly develop and deliver policy aims. The traditional approach to government work in the United Kingdom or Canada is for policy teams to develop policy and legislation.

They go through a formal consultation period, Parliament debates, legislation is set out and handed over to an operational team to deliver, and probably a separate IT team develops a website.

The Ministry of Justice in the United Kingdom employs almost 80,000 people. It supports and runs a legal system that's existed and evolved in various forms for a millennium, and is responsible for running courts, tribunals, and prisons across the country.

After the creation of GDS in 2011, the Ministry of Justice was an early and enthusiastic adopter of new ways to deliver services. It started by introducing a simple way for families to book visits to see their loved ones in prison, which saved people days of uncertainty and hours of wasted travel time and helped support rehabilitation.

Other services that it's redesigned, such as for people eligible for financial help with court fees, addressed the needs of internal users, too.[6] The teams recognized that the staff using the systems had unmet needs that were slowing them down and driving them to paper-based approaches.

The Ministry of Justice is now one of the leading U.K. departments taking the digital approach further to actively teach policy-makers different ways of working. This means policy is developed by talking to the people it impacts directly and using this approach to understand how effective the policy will be in achieving the government's aims. The U.K. government has traditionally consulted on policy, and there are often representative bodies and stakeholder groups closely involved, but it's less common to frequently engage with users.

The U.K. Experience of Setting the Bar for New Approaches

After the GDS team had quickly built and iterated the initial version of gov.uk, and the migration of all previous government websites had begun, attention turned to how this knowledge could be shared and how these approaches could be embedded across the government. This was a two-pronged approach. One of the important mantras of GDS was "the strategy is delivery." The best way to demonstrate the art of the possible was to get on and do it while others were busy looking for reasons not to. So, building on the creation of a user-centred, single-government website in gov.uk, GDS worked with departments to identify key services that needed to be redesigned for digital delivery and then sent in people to support that. This was largely successful and resulted in 20 redesigned services being made available in 400 days.[7]

The main point of the "exemplar" program, though, was not to make 25 services better but to show departments how empowered, multi-disciplinary teams expected and needed to work, so that when the GDS team withdrew, the culture remained.

The Department of Work and Pensions set up its own digital academy to train its staff in new ways of working, and that's been incorporated into GDS and a broader training curriculum for digital practices across the government.

ENSURING BEST PRACTICE IS FOLLOWED

Setting principles and having clarity on common approaches to good practice are important to rally people who already hope to go in the same direction. But for those who don't agree or whose priorities are different, principles will largely be ignored unless they're relevant and enforced.

The Importance of Setting Standards

Alongside the ability to go in and work with departments on an invited basis, GDS had two mechanisms to enforce change across the government, which were two of the important differences between the U.K. and U.S. approaches to digital. The Government Digital Strategy published in 2012 gave GDS the first of these mechanisms — the ability to set and enforce standards for digital development.[8] In setting these standards, GDS set out how a team delivering a digital service should operate. These principles were iterated and adapted, but most of the fundamental ideas remained the same.

Alongside supporting departments to deliver services differently, GDS presented the standard for what the teams developing these were aiming for. In 2013 GDS began sharing a Digital Service Standard, giving departments time to adapt and prepare. In April 2014, this was made mandatory for all new or redesigned public-facing digital services.

The Digital Service Standard comprised the knowledge of those who set up GDS and reflected what they had learned in building gov.uk. The first iteration was recognized as a draft, pulled together quickly from the knowledge and skills of those involved. But it gave everyone somewhere to start from. And for the first time, across government, GDS was able to say "to develop a good digital service you must do it in this way."

GDS's role in enforcing these standards was sometimes controversial. The standards required departments and teams to operate in a very different manner and to bring in new skills. All new and redesigned services had to be assessed by a GDS panel at three stages of development: the end of their alpha stage, before the service was made public as a beta version, and before the service was deemed fully live and beta branding was removed.

When a service didn't meet the standard, it was given a set of recommendations, which could sometimes be to return to discovery phase, and it wasn't given a gov.uk domain, meaning the service couldn't be made public. The commitment to transparency was quickly demonstrated by the publication of all assessment reports — both those that met the standard and those that didn't — and the recommendations for improvement alongside them.

As the first few months of implementation of the service standard progressed, GDS and departments started to learn more about the way this worked in practice. The standard was iterated and the number of criteria for digital services reduced. And as other countries and states have adopted a similar approach, they have been able to learn from the United Kingdom and leapfrog to add their own takes on important issues.

Australia condensed and tightened language, while the Government of Ontario in Canada chose to emphasize the creation of the right team. In the Canadian federal government, the principles of the Digital Service Standard were combined with those of the U.K. technology code of practice, and included a principle on ethical use of technology, recognizing the important role AI will play in future service delivery.

The U.K. version is currently evolving to set the bar higher again and reflect the nature of full end-to-end user journeys, including elements that aren't on the Internet.

As departmental teams and capabilities have grown, the assessment model has changed and assessors are volunteering from across government departments to assure one another's work, with GDS holding a quality-control, standard-setting, and facilitative approach.

Gaining Financial Control

The second important mechanism the U.K. government introduced and ended up moving to within GDS was spending controls for digital and technology costs. In the wake of the economic crisis, the minister for the Cabinet Office, the then Right Honourable Francis Maude, MP, set out strong central controls across a number of areas of government spending, of which IT was one.

In 2014 the control over digital and technology spending was linked to a set of expenditure red lines[9] that were supported by the technology code of practice. Both of these were intended to fundamentally change the way the government spent money on technology; to end monolithic, long-term, single-supplier contracts; and to move government toward a user-centred, agile, iterative approach, making sensible use of commercial products where they were appropriate. The red lines originally set out were the following:

- No IT contracts over £100 million in value unless there is an exceptional reason to do so.
- If a company has a contract for service provision, it shouldn't also do the service integration for that service.
- No automatic contract extensions.
- New hosting contracts to last no more than two years.

And over three years, at a conservative estimate by the National Audit Office, these controls saved the U.K. government £1.3 billion. More important, they supported the other work of GDS by requiring departments to rebuild in-house knowledge and expertise and to move IT from procurement oversight to technology development.

KEEPING PACE WITH RAISED EXPECTATIONS

The ability of governments to deliver services in a manner people expect, and to remain relevant, depends on their capacity to understand and keep up with the expectations of citizens.

What Can Government Services Look Like in 50 Years' Time?

As the youngest generation turns 18 and starts to directly engage with government more and more, governments need to respond to the expectations of these young people who have never lived without the Internet. They also need to focus on areas where good services offered by private service providers are allowing people with disabilities, or those who are unable to travel, increased autonomy and the opportunity to engage.

There's a very short answer to the question "What can government services look like in 50 years' time?" And that is: "They can be invisible."

The real question is whether that's a political choice governments want to make. There's no technology-based reason why services need to have citizens or businesses engage with government in order to deliver what people need. But given that sometimes we value what we see, even when it means a slightly less efficient process, will governments want to remove that altogether?

In a world where millions of people are asking algorithms on dating websites to choose the person with whom they should spend the rest of their lives, governments must be able to make use of technology to allow people simple, straightforward access to the services they're entitled to and allow them to fulfill their responsibilities as citizens, securely, and with minimum effort.

Let's look at an example. Cash transactions will be largely, if not fully, non-existent in the next 10 years. Depending on the approach of a government and the appetite of citizens for personal privacy, governments already have access to all the information they need to accurately calculate tax, which is already common practice in some countries. Using AI, governments could make calculations to a high degree of accuracy on any areas where data isn't held. But governments might choose to show their workings and might be keen to demonstrate through this their relevance to the lives of citizens.

Global Citizens and Virtual Citizenship – National to Borderless

The way people live has changed significantly in the past 50 years, giving rise to expectations that governments have had to adapt to meet. As people continue to travel more and the demand for new skills increases worldwide, forward-thinking countries are adapting to welcome this change and use it to build their own economies.

Estonia is a small country of 1.3 million people, bordering Latvia, Russia, and the Baltic Sea. Forest covers 48 percent of the land. Estonia has chosen over the past two decades to use what some might see as its weaknesses to build its competitive advantage over other nations as a world leader in digital government.

Estonia's path to efficient digital government has been significantly aided by a national citizen ID, something that has been politically unpalatable in Canada, the United Kingdom, New Zealand, and others. The ID card was introduced in 2001 and gives Estonians secure and authenticated access to all government services online. A national citizen ID, combined with acceptance of data-sharing across government departments and the private sector, has also been a key factor in better delivery of government services in Estonia.

Despite having achieved that significant service improvement, Estonians have continued to push the boundaries in their drive to create a digital nation. In 2014 the country set up an "e-Residency" program. Through this program, people around the world are able to become e-Residents of Estonia and start a business there. For each new e-Resident, the country gains income through the use of Estonian businesses and services. It's proving popular among people in nations such as Ukraine where transferring money in and out of the country is difficult, and increasingly in the United Kingdom where people would like to have a European Union–based company after Brexit is fully implemented.

Estonia has also created the world's first "data embassy." In an agreement it's made with Luxembourg, the Estonian government has established computer servers in that country that it owns and controls outside its territory. As data becomes an increasingly valuable commodity, if a country can effectively operate from outside its geographical boundaries, what does this mean for the future of the nation-state?

Governments and industries have yet to move forward to address basic needs for people who move from nation to nation. For example, an individual's credit history doesn't transfer when he or she relocates internationally, which means it has to be built from scratch. Countries that are able to get ahead on issues such as this are the ones that are going to attract the best talent.

Although currently there's very little being done to actively address this situation, the right signals are being sent. The presidency of the Council of the European Union is held on a rotating basis and changes every six months to a new member state. For the second half of 2017, Estonia held the presidency, and as is customary for the president, was able to set the themes for discussion.

Unsurprisingly, Estonia chose to focus the attention of member states on digital Europe and the free flow of data. In the early weeks of its presidency, it was successful in delivering a unanimous declaration from all members to work toward user-focused, cross-border services, with a "tell-us-once" principle. A successful "tell-us-once" service principle that allows people to move seamlessly across borders while their data moves with or ahead of them is a long way off. But, if achieved, it would significantly strengthen the European Union's position as a location able to attract the most talented people.

TO LEAD, TO FOLLOW, OR TO BE LEFT BEHIND?

For countries like Canada, there is a choice to be made: to lead, to follow, or to be left behind? To embrace the changing expectations of its population or to protect the status quo of how government has always operated? Governments across the world, including Canada's, have demonstrated time and time again their ability to adapt and innovate. Building on the experience of other countries, the expertise of 17,000 public servants working in the IT domain, and a thriving AI research sector, Canada is well placed to lead and become a world-leading digital government.

*　　*　　*

ESSENTIAL TAKEAWAYS

Olivia Neal cuts to the core of why governments need to be digital. All our services must be designed for the digital world, and simply digitizing analogue services won't achieve that. We must completely re-engineer how we do government services, and this chapter is loaded with important lessons learned in this area. Like many of the contributors to this book, she bears the scars of having gone through the digital journey.

1. *Government service is no longer about trans-actions.* It's about seamless service. How does a government deliver a service on any platform (Facebook, travel sites, et cetera) and on any device (refrigerators, watches, cars, et cetera)? How do governments allow third parties to assist in the delivery of services in the age of platform economies?

2. *Services must be delivered where the client wants them.* During a time of digital accessibility, it's disrespectful for governments to force citizens to come to them. This is a powerful shift requiring a massive cultural change in most public-sector institutions.

3. *Government culture needs a reset in order to embrace the digital age.* As public servants, we must modify our mindsets. We must challenge the analogue way of thinking that has permeated our institutions, defying the status quo every day. Not because we're bad civil servants but rather because the viability and relevance of the institutions we love are in jeopardy if we don't. Our citizens and businesses are digital, and in many cases, our governments aren't. The market is forcing us to change, and we must embrace this reality.

NOTES

1. Tom Loosemore (@tomskitomski), "Digital: Applying the culture, practices, processes & technologies of the Internet-era to respond to people's raised expectations," Twitter, May 10, 2016, 3:00 a.m., https://twitter.com/tomskitomski/status/729974444794494976.

2. "ICT Facts and Figures 2017," ITU, accessed April 7, 2018, www.itu.int/en/ITU-D/Statistics/Pages/facts/default.aspx.

3. "Performance,"GOV.UK, accessed April 7, 2018, www.gov.uk/performance.

4. "The Quality of Service for Personal Taxpayers," National Audit Office, May 25, 2016, www.nao.org.uk/report/the-quality-of-service-for-personal-taxpayers.

5. Richard C., "NCSC IT: How the NCSC Built Its Own IT System," *NCSC* (blog), GOV.UK, January 29, 2018, www.ncsc.gov.uk/blog-post/ncsc-it-how-ncsc-built-its-own-it-system-0.

6. Leigh Money, "Making It Easier for Staff to Process Help with Court Fees Applications," *MOJ Digital & Technology* (blog), GOV.UK, January 5, 2016, mojdigital.blog.gov.uk/2016/01/05/making-it-easier-for-staff-to-process-help-with-court-fees-applications.

7. Mike Beaven, "Looking Back at the Exemplars," *Government Digital Services* (blog), GOV.UK, March 27, 2015, https://gds.blog.gov.uk/2015/03/27/looking-back-at-the-exemplars.

8. "Government Digital Strategy," GOV.UK, December 10, 2013, www.gov.uk/government/publications/government-digital-strategy.

9. "Government Draws the Line on Bloated and Wasteful IT Contracts," GOV.UK, January 24, 2014, www.gov.uk/government/news/government-draws-the-line-on-bloated-and-wasteful-it-contracts.

8

Building a Digital Government the e-Estonian Way

Siim Sikkut

I MET SIIM SIKKUT for the first time in 2017 in Ottawa at the inaugural Forward50 (FWD50), a conference designed to bring digital government thinking from around the world to the Canadian public sector. I fell in love with his vision. A decade ago government was big, needed to do business with big enterprises, and only big mattered. Estonia is small, agile, nimble, and capable. It has developed a vision for a virtual state that is unmatched anywhere in the world. Much of this is due to the work Siim has done in his country. He and the team in Estonia have showed the world that when governments go digital big is irrelevant. There is also a major difference between digitizing analogue government and building digital government. When shifting to a digital world, policies, strategies, programs, and services must be changed. Creating an ecosystem of government that ensures multi-channel service delivery becomes more important. Making government services ubiquitous is what truly matters. Siim steers us through what it means to be an e-Resident in Estonia – a model that the rest of the world needs to understand, implement, and evolve. Imagine

this: could we be on the verge of finally breaking down physical borders? Okay ... okay ... perhaps not quite yet, though
Estonia and Siim might beg to differ.

— Alex Benay

We started calling my country, the Republic of Estonia,
by the nickname e-Estonia to highlight the great extent to which digital services and solutions have become a part of our daily lives — as residents, as entrepreneurs, and
as officials and leaders in government.

After 20 years of work, digital has become our way of life. It all
started with the banking sector where Estonia skipped the whole
chequebook phase and jumped straight to payment cards and online banking in the mid-1990s. Government came next with the
digitization of agency back-office processes, while also bringing
bureaucratic front-line services online, one after another, just as
the Internet started reaching the masses in the late 1990s.

DIGITAL GOVERNMENT IS A BETTER PERFORMING GOVERNMENT

The benefits of our digital journey are very evident. They can be
seen on the aggregate, or macro, level where the whole government
now performs better in terms of service delivery, quality of service,
and efficiency. But, even more powerfully, the digital benefits are
there in the value that digital government brings to people's daily
lives and businesses. There are real-life examples of this from all
kinds of walks of life.

Ninety-six percent of Estonian residents have submitted personal tax declarations online since the early 2000s. In a sense,
we're actually a funny country because most people don't mind

paying taxes now that it's so simple to do. Save for a tiny fraction of people (e.g., those with foreign incomes), paying annual taxes takes fewer than five minutes: you log in, review the data, click "next-next-next," and then are done! There are no payment slips to hold on to throughout the year, no tax accountants (the profession doesn't really exist in Estonia except for corporate clients), and no other costs. Of course, this is all because of the general setup of the Estonian tax system — it's simple and transparent.

Since 2007, Estonia has been working on a nationwide electronic health record system that can generate digital prescriptions. Doctors can access the medical histories of their patients or consult one another digitally from their desks, resulting in quicker diagnoses and likely more accurate ones. People can also get their medicine without visiting doctors, since prescriptions can be issued digitally and picked up at pharmacies right away. Thus, there's no need for paper records or unnecessarily duplicate doctor visits. This is a big efficiency gain, since in Estonia, as in most developed countries, many people who visit doctors for prescriptions suffer chronic conditions that don't usually require re-examinations. Therefore, this time saved can now be used by doctors to see patients with more acute needs.

When it's election time, Estonians can vote on the Internet. The country was the first to introduce Internet voting and still remains the only nation globally to have it as a viable option. Contrary to what so-called experts say who pop up for politically motivated causes every now and then, the procedure is fully secure, more so even than traditional ways of voting. For example, for Estonians, postal voting sounds completely ludicrous! There are so many ways votes sent by mail can be intercepted, altered, destroyed, or otherwise tampered with that require far less effort or money than compromising the online voting platform.

Estonia has already had nine elections for the national and European parliaments, as well as for municipal elections, and not a single incident has occurred with online voting. Also, in autumn 2017, Estonia had another round of municipal elections in which a third of the votes were cast online by Estonian residents from 115 countries around the world. So online voting makes democracy and civic participation more accessible and secure in our global, ever busier, and increasingly interconnected time and age. Estonian citizens can vote from Pearson International Airport in Toronto, from a beach in Fiji, or simply from their office desks or on couches in their homes. Because we've made voting so convenient and accessible, we haven't had the decrease in overall voter turnout that many countries are experiencing.

Today digital self-service and interaction options exist for virtually all kinds of dealings between the Estonian government and its citizens or enterprises, at least on the national level. It isn't so in every single municipality yet, but there are only very few service functions on the municipal level in our small country. The only government services in which face-to-face interaction is still required are buying and selling real estate and getting married or divorced. These services represent some of the biggest decisions people need to make in their lives, so we want to ensure they're made voluntarily and with full awareness, even if all the proceedings of notary or marriage registration happen digitally onsite. People also have to attend in person to pick up identity documents, though work is under way to make that accessible through a variety of service providers.

For entrepreneurs, not only are all government services available online, but it's actually mandatory to transact digitally with the public sector after a company has been registered. A firm can be registered using a paper form (i.e., through a notary) but very few do so. Nowadays it takes on average between a few hours to one day after the online application has been submitted to have a company legally set up and running. Actually, our current record for the fastest company to be set up is 18 minutes! All the necessary controls and checks are done digitally within that time by public authorities, which means a business can be operational very quickly with legitimate corporate powers. We made this possible to encourage entrepreneurship and business start-up activity in the country.

All the other things new businesses have to take care of such as reporting, tax declarations, et cetera, are handled digitally, as well. Entrepreneurs don't have any need to wait in public offices to submit official documents to the government. Thus, instead of spending time on red tape and bureaucracy, they can focus on what matters the most — growing their businesses. A business-friendly environment and procedures reduce hassle, which directly leads to lowered costs of doing business. This translates to productivity gains and profits, especially for small- and medium-size enterprises (SMEs). On a macro level, this results in more economic growth for the nation.

In addition to creating a better business environment, digital solutions support growth through efficiency in public-service delivery and more effective governance. This was the very reason why Estonia started going digital two decades ago at the height of the Russian economic crisis of 1998, which created a financial downturn and a sharp fall in tax incomes. So Estonian politicians searched for any and all efficiency solutions and took the advice of techies to begin implementing digital technology throughout the public sector.

As it turns out, in digital ways we indeed were — and still are — able to deliver services for better value and less money, both in the back office and front line of public administration. Based on initial experiments, the digital path became the strategy to follow and still is.

Digital government has allowed us to be very lean, contributing greatly to fiscal prudence, which Estonia takes pride in and is known for. The country has the lowest sovereign debt burden in the European Union, less than 10 percent of gross domestic product (GDP) now. We also see the effectiveness and efficiency gains in every policy sector. For example, our tax system has become the most cost-effective one in the Organisation for

Economic Co-operation and Development (OECD), based on amount of tax revenue collected per euro spent on tax administration. This has allowed several consecutive governments to keep reducing personal income taxes over many years. In a similar vein, our health system is very efficient with very little resources needed despite the growing aging population. Our Estonian police departments are able to solve more crimes using digital tools and data. This list of examples goes on and on. In short, digital government means a government that can better deliver its services and policies for the people.

FOUNDATIONS FOR SUCCESS

Estonian ministers and officials are often asked: "So how did you do it? Can other countries do the same?" On the surface, the Estonian experience is often discounted. Yes, we're a small nation of only 1.3 million people with a governance setup that's very lean. This does make decision-making faster, since there are fewer people who need to be convinced. Yet, even in big countries, quick decisions can be made if resolve is put into the process, as we have seen.

We surely also benefited from having no significant technological legacy or infrastructure when we started down the digital path. Instead of mainframes and COBOL, we were able to go straight to the modern World Wide Web and Java as we rebuilt the country and restored our independence after the collapse of the Soviet Union. Yet all countries have legacies in an organizational, legal, and cultural sense, and Estonia was no exception. To get to the efficient digital services, we had to break down the existing habits and processes by reforming policies, simplifying procedures, and redrafting legislation. Breaking the technological legacy is only one, and in our view, lesser aspect of the change necessary to make a government digital.

1. Leadership and Political Will

This brings me to the first lesson we learned on our journey to digital government, and the first pillar, or requirement, that needs to be in place for success in digitalization: strong leadership, both on political and administrative levels within government. Technology is merely an instrument that enables change in specific, effective ways, but it's the change itself that needs to be managed. The buzz term is "digital transformation" for a reason. Technology helps simplify and reform but alone doesn't do the trick. Estonia was able to make tax declaration or company registration quick and easy because we redesigned or reinvented the processes behind them, not just automated or digitized the complicated bureaucratic process as it was set up before.

Resolving to move fast enough and break legacies demands political will to push the agenda through and to do things differently. It also requires acceptance of risks that always come with innovation. The initiatives can and will fail sometimes, and they will need to be iterated; they won't be perfect from Day 1. But leaders need to realize that this is completely normal in the world of technology, as we see with many start-ups. In fact, I would argue that in our fast-changing and globally competitive world, it's the political will and (calculated) risk-taking that will be the biggest make-or-break factor to determine a nation's prosperity in the years to come — not just in succeeding to build an effective digital government success but in other endeavours, too.

2. Shared Service Platforms Especially for Digital Identity and Data Sharing

The second prerequisite, or at the very least a very powerful enabler for building an effective digital government, is to put in place the proper technical architecture and infrastructure — the digital service platforms that are used and shared across government. In Estonia, as we started the digital transformation, we didn't have enough money in the government coffers for all our IT investment requests. This meant we had to optimize how we did things, so we came up with the idea to assemble shared platforms for recurrent functionalities from potential IT projects. Basically, we created shared solutions and a common service infrastructure, since we couldn't afford building up everything in silos and in duplicate. As an additional bonus, it made digital transformation for various processes or services much swifter, too. Afterward, there was less need to reconstruct the same functionalities, and more effort could be spent on the actual redesign and digitizing of actual business processes of government. This meant we avoided duplicating our efforts and resources from agency to agency to understand the underlying technology.

The most notable example of an infrastructure platform we moved to was the nationwide digital identity. In Estonia every resident has to have a national ID in the form of a smart-chip-based ID card, which also acts as physical identification. Through a collaboration with the government and telecom providers, Estonian citizens can receive a special SIM card for mobile phones called a mobileID, with the same functionalities as the ID card. In order to utilize them, unique, secret PIN codes are needed, similar to the kind used with a bank card.

All public digital services allow, or even require, the national digital ID for entry and use, whether they're accessed by public servants, citizens, or company representatives (that is, all parties involved in the interaction). This means that, as agencies digitize their workflows, they don't have to reinvent user authentication each time. They can just reuse the common components already built and provided centrally by the domain of the government chief information officer (GCIO).

In addition to user authentication, the Estonian digital ID allows users to sign anything and everything digitally. Since 2002, digital signatures with national IDs have been recognized as legally equal to handwritten signatures. Because physical signatures are so easy to falsify, especially in the digital age, digital signing has actually become more secure. The benefit of digital signatures is that each and every interaction can be made fully digital and moved online. Quite often in the world we see services that are digitized beautifully until some legally consequential act of authorization has to occur. Then, at the very last stage, documents still have to get printed and mailed, or a trip to a government office still has to happen to provide the signature. But not in Estonia! With digital signatures, we've been able to move bureaucracy online completely.

Digital IDs and signatures don't only benefit government operations, though. In fact, their biggest usage and benefits arise in the private sector. The same national ID can be employed to log in to an online bank account, movie theatre account (as a loyalty card), or utilities providers' online self-service sites. Because of digital signatures, enterprises don't have to spend money or time for trips and meetings to make deals with other companies or hire new employees. Instead, they sign contracts digitally, wherever they are in the world, and can do so within seconds rather than wait days or weeks for the deals to be completed.

In fact, most digital signatures are given in business-to-business or business-to-customer interactions. So digital IDs and signatures enable not just the digitization of public-sector services but the entire economy. The platforms we built for the public sector have indeed become the national infrastructure by being open and usable for all, very much in accordance with the "government-as-a-platform" model.

The value of digital IDs and signatures is economically significant. We estimate, in the back-of-the-envelope-calculation way, that Estonia gets at least 2 percent of GDP worth of efficiency gains each year because of digital signatures alone. This is based on multiplying the number of digital signatures processed annually by a conservative estimate of time that it would take to sign a document (travel, waiting, process, et cetera) and convert this time saved to working weeks. We save at least one work week for every employed person in Estonia, and there are about 50 effective working weeks in each year, i.e., the time it takes to produce a year's worth of GDP. This shows there's a real economic benefit and growth contribution from digital solutions. Of course, not all of this time saved will go directly to producing additional economic output. It can instead mean more time for hobbies, families, and so on. And that's okay, because that leisure time can potentially contribute to better productivity in the future.

Another key universal challenge across any IT projects and for all governments is integration — interoperability and data sharing, specifically. In Estonia we solved this issue through a common data exchange platform called X-Road, which allows us to transmit and exchange any data in a highly secure way between any organization and information system. Once a system is made compatible with X-Road, it's immediately interoperable and integrated with all other X-Road nodes or connected systems. Thus, a one-time effort provides immediate government-wide interoperability to send or receive any data that an agency might need.

X-Road, in a nutshell, is a very lean, standardized exchange protocol, similar to a standardized application programming interface (API), in a sense, and is the interoperability layer on top of all technical systems that exist in government agencies. It isn't a central repository for all the data the government holds, nor is it a super database or a central service bus. Instead, all exchanges happen directly between systems and don't flow through any central components. As such, the platform allows for effective resilience against hacks or service disruptions. Even if some systems are down, the rest can continue operating fully and safely, and can still exchange data with the rest of the government.

A data exchange platform can greatly enable digitization, since user needs should lead us to build and deliver services to people in an integrated manner. Citizens and companies don't care about government silos. They want to get things done in single encounters as much as possible, and as fast as possible. For example, the reason why our tax declarations are very fast is that we offer them prefilled by digitally pulling together pre-collected data from different systems via X-Road. The outcome is that the service creates much less hassle and works seamlessly.

The Estonian X-Road has also become something of a national infrastructure, since it's used and is applicable beyond governmental processes and services. Private-sector entities have been integrated with public databases and agencies for a long time, so they can send or receive data from the public sector or ensure seamless delegation or reporting with it. In addition, the platform is open and usable for other entities for their data-sharing needs. For example, all Estonian smart grid operations and exchanges in the electricity system are built on the platform and utilized by all (private and state-owned) energy producers and distributors. Currently, we're beginning to explore how to implement X-Road in the growing Internet of Things (IoT), potentially iterating it into a platform that allows secure data transmission and interoperability across value chains in the economy. The platform is also open and usable for other countries, and Finland and Mexico, among others, have now adopted X-Road to be the infrastructure of their digital governments.

3. Enabling Legal Framework

Some of the most important infrastructure in a country is not technical but legal. Suitable rules and regulations make up the third prerequisite for unleashing digital transformation.

At the very least, the law can't be a barrier to digital transformation. Quite often, digital innovation is held back by notions that "the law doesn't allow that." But laws can be changed and, in fact, have to be changed to make the most of digital means. There has to be willingness and effort to rewrite existing rules if they don't make sense anymore in a digital world. The simplest (and most mind-blowing) example can be to get rid of explicit requirements that some processes and documentation need to be paper-based. A successful digital redesign of service always requires a redesign of underlying rules, since that's where the bureaucratic service processes are enshrined.

As a side note, in my experience in Estonia and around the world, quite often the issue is not even the letter of law but its limited and outdated interpretation. This is why government CIO offices and digital teams often have lawyers of their own to challenge the status quo faster and more effectively by figuring out and explaining the possibility of alternative interpretations. But, at the end of the day, the cleanest and surest option to break this barrier is to get rid of outdated legislation and regulations or to bring in new and clearly enabling ones.

In Estonia's experience, the legal framework can actually be reconfigured to assist digitization more powerfully than just removing the barriers. We've tried to use legal rules as enablers and push-factors for change. For example, years ago we made into law a principle called once-only, which mandates that government agencies shouldn't ask again for data that some part of government already holds. People and companies can refuse to resubmit it. This forces service redesign in a seamless, integrated direction, raising user expectations and the standard for service delivery. On a technical level, the rule compels us to make great strides in interoperability and solve the often endless technical arguing on how to link systems. This is why we brought in the principle in the first place — to make sure X-Road was used after its uptake had only been moderately successful in the early years when it was in voluntary mode.

4. Managing the Risks

Having said the above, certain legal barriers are essential as safeguards. The fourth pillar for digital success is to take seriously the risks digital solutions can bring.

Our e-Estonia is very much built on trust. Without trust, people wouldn't use public digital services. Without users, we wouldn't be able to provide these services and reap the benefits they offer to the extent we do.

Trust starts from the legal framework. Strong data protection is essential for trust to be there. However, "strong data protection" isn't another way of saying "you can do only very little with data." In Estonia we have some of the strongest data protection frameworks in the world, since we're a member country of the European Union. But that hasn't prevented us from bringing in the principle of once-only, or reusing and exchanging data in the public sector between different agencies. Strong data protection means we have to have justified reasons to do so (there has to be a "need to know" to access data on both agency and individual levels, which we can manage through access controls), and once we hold or process data, we have to do so safely.

The duty to ensure that the rules are enforced lies first and foremost with the agencies handling the data. However, we've gone beyond that to ensure public trust in data sharing. Our laws are built on the notion that it's the people and companies who always own their personal data, not the government. One of the implications of this is that we have to let people see how the government utilizes their data. The classic way to do so is upon explicit written request, which is an often slow and taxing bureaucratic process in itself. Yet digital technology offers a radical alternative: why not proactively open up the logs to people themselves in real time? That's why, in Estonia, people or companies can see which official or agency looked up their data, and when, through government portals any time they want with just a few clicks. In this regard, the people act as "Big Brother" over government, and not the other, often-feared way around. If citizens suspect that an official or agency had no business to look at their particular data, they can make a claim to the data protection office and an investigation will

be launched. If an unsanctioned action or a violation of privacy is discovered, significant punishments follow. Thus, transparency is the ultimate source of trust in Estonia.

Also, not only can people have easy oversight of how their data is used but they can also control it directly to truly exercise their privacy rights. For example, when it comes to online health records, citizens can log in and block access to some or all of their digital health history if they don't want it to be available to doctors. People do want to manage their personal data, and that's a function I believe all governments should offer in support of fundamental human rights, but also to ensure citizen trust and gain support for digital changes.

Another aspect of trust is data security so that personal data doesn't get stolen or altered without authorization and our systems are always fully operational when needed. Due to our high reliance on digital solutions in Estonia, cybersecurity is a critical function for us. We work hard to maintain it, since we don't want to go back to analogue days. Basically, cybersecurity is the daily "hygiene" needed to be digital in the manner we want to be.

Our cybersecurity strategy in Estonia is multi-layered. It all starts with how we design the services and systems in the first place — by figuring out how to set up processes in a way to best avoid potential misuse and risks, using state-of-the-art technology (e.g., the newest cryptography), and so on. We also spend a lot of time raising awareness among both public servants and the general public — our end-users — on how to avoid cyber threats. Of course, we also have to be on guard and keep our defences up because, regardless of how well a system is designed and how aware the users are, no system can ever be 100 percent secure. This is certainly the case if systems are specifically targeted, especially in a world where nation-states are becoming aggressively active in cyberspace. Therefore, Estonia invests increasingly in its capacity to detect incidents, manage them, and learn from them. Our cyber defences have been well battle-tested, such as in 2007 when Estonia was the target of the world's first cyber attack against a country and managed to defend itself successfully. The attacks continue to occur literally every minute, but we've gone through them without any major hiccups. Yet it's always better to be safe than sorry, and such work never ends.

The message here is that cybersecurity and trust, more widely, better be part of a national digital strategy for the latter to be sustainable and effective. Hopefully, Estonia's experience also refutes the often heard claim that the risks digital solutions bring are simply too large to go fully digitally. In the Estonian view and experience, privacy and cybersecurity don't have to be barriers to digital progress. They're risks we can manage to an acceptable level by working on them. That way cybersecurity and privacy are really enablers, making the progress of digital society possible. They shouldn't remain as excuses for not going digital.

FUTURE OF DIGITAL SERVICES: INVISIBLE

As technology develops and the horizon of our imaginations extends, the work to build digital governments will never be done. The same goes for Estonia. Even if nearly all our services are digital, we can continue to reinvent them to be even better.

For me and my teams, the next goal is to make service delivery proactive, and further to that, government services invisible. The best user experience is one in which the users don't need to engage with government and bureaucracy at all, and their needs are met right when they appear, or better yet before, they arise. In Estonia we have all the underlying elements needed in place for this, such as the world's most extensive data exchange capability and strong digital identity. We just need to add some more automation, better integration, analytics, and a whole lot more redesign.

Most service needs appear in relation to specific events in the lives of citizens and businesses. Milestones occur in people's lives, such as a marriage or a first-time purchase of a house, and they need to get a lot of things done all at once. Other than that, people rarely interact with the government as actual evidence shows. This is what we are focusing on in our redesign — making services proactive around the life events of citizens and businesses, and redesigning services to require, at most, only a single encounter with government. The departments, agencies, and other actors should collaborate to deliver services behind the scenes and remain there from the user point of view.

For example, when a business is started, all the necessary paperwork should be processed immediately without requiring further registration of employees or licence approvals. When a child is born, the government should be able to send a congratulatory e-mail with a link to submit the minimum amount of data actually needed to set the necessary actions in motion. Through a simple online form, new parents should answer the following questions only and at the same time: What will be the name of the child? In which bank account should child benefits be transferred to (we probably already have that on file from tax returns, so we only need confirmation)? In which kindergarten should the child be registered in (we can suggest the three nearest ones to the registered home)? And these questions can be answered online from a hospital bed before leaving for home with the newborn. But right now there are five different applications on five different websites with five different user experiences to get the same thing done, but even these can be done the same day from a hospital bed, too. We definitely could improve all five of them, or we could take a bigger step to integrate them into a single interaction and really provide convenience and efficiency for users.

Some interactions can be fully automated right away. For example, we're now building automated reporting for businesses. If they consent to giving our government tax office direct access to their financial data (e.g., in accounting service or software), the companies will never have to submit any tax declarations, statistics, annual reports, or other such documents and data to the government ever again. Basically, by connecting directly to the originating point of the data (which is often already digital), the government can automate the paperwork processes for entrepreneurs. Companies would no longer necessarily even need accountants and could focus simply on doing business if their financial flows are straightforward enough. For micro-companies and SMEs, this is usually the case, and they could get a radical efficiency boost this way.

This is our vision as a government — to offer a zero-bureaucracy environment as often as possible using digital tools that are already available. Artificial intelligence and big-data analytics can make such proactive work more powerful, even leading to predictive policy decisions. However, one doesn't need complicated data science to get started in this direction. Often the reuse of existing data and tools is enough. In the case of registering a child, as outlined above, the government already knows that the child was born, since the hospital makes an entry into the population registry (before the child even has a name). A proactive interaction can already be triggered from this single data entry.

COUNTRY AS A SERVICE

In the view of Estonia, another future direction is public digital service delivery that's usable across borders. That's already happening regionally, especially in the European Union, in the context of the Digital Single Market where a lot of effort is being put in to legal harmonization, intergovernmental interoperability of services and platforms, and even pan-European platforms.

This direction may lead to more competition between governments on the merits of their digital services. Countries already compete to be the locations of choice for investments, talent, and company offices. That's often based on tax arrangements and local production factors. But in the age of the digital economy, the crucial factor will be the ease of doing business online.

That's why we started the e-Residency program at the end of 2014. Estonia is one of the first nations in the world to make its services available for non-residents around the globe. Anyone in the world can become an e-Resident of Estonia by applying for our state-issued digital identity, and getting authentication access and digital signature capability to use the services offered online by the Estonian government and private partners. For the latter, e-Residency is available as a platform for onboarding new customers or authenticating existing ones, e.g., for financial service providers.

The core target group of e-Residency is cross-border entrepreneurs, who can now run their transnational business virtually through Estonia using its digital ID and signature services in a trusted way — with very little hassle and cost, no middlemen, and no location barriers. It's very useful for the growing number of freelancers and micro-entrepreneurs around the world who may do online e-commerce or provide their professional services remotely. And it's especially useful for those who want to be very mobile in their businesses and live and travel the world while working.

Interestingly, we also see people from around the world applying to become e-Residents because they don't have convenient digital services available in their own locations or home countries. So, essentially, Estonia is their country-as-a-service choice to get some public services or goods. People physically reside elsewhere but make use of the Estonian jurisdiction and hassle-free digital offerings.

So far Estonia is the first and only nation with such a value proposition. But we expect several other countries with strong enough digital identity programs to follow suit fairly soon, since this is a lucrative area to pursue. For governments that are slow to digitize, this could be a warning sign that other nations will start serving their jurisdictions and perhaps do a better job. The result could be an exodus of talent, a decrease in tax revenue, and a drop in economic growth unless countries start delivering services to their citizens in a digital and user-friendly manner. At the very least, nations will suffer from lack of confidence or dissatisfaction from their citizens if they don't advance comparatively in the digital sphere vis-à-vis other jurisdictions.

In Estonia we are already getting more e-Residents each week than babies born in our country. Thus, we grow faster as a digital nation and a digital economy these days than in the traditional, physical manner through population growth. This is exactly why we started the program: to make the Estonian economy bigger without having to rely on the growth of our population, which takes years. We connect new users with our digital service providers and allow them to grow faster than they otherwise could in a small domestic market. Thus, digital government allows us to break out of our physical constraints.

To briefly summarize, the e-Estonia story shows that digital development can reap huge benefits. It pays greatly to digitally transform, and to do so fast. I hope that Estonia is living proof that a digital government can indeed be built despite the many challenges. Digital transformation is possible; it just takes a bit of courage and a lot of work. But it can be done and should be done.

*   *   *

ESSENTIAL TAKEAWAYS

The notion of "country-as-a-service" may be a bridge too far for many senior public-sector executives and political leaders. But the reality is that it has been done. Siim Sikkut shows us how this can be accomplished and why we all must listen. What's impressive is that Estonia transformed itself with rigid planning, disciplined execution, and revamped standards and architectural approaches to technology management. In a way, the vision is extremely ambitious, but the execution is simple in its approach. The juxtaposition of these two elements makes the Estonian model a stroke of genius. Here are some key takeaways:

1. *Political leadership is a must for such a transformation.* Legislation can often be a barrier to digital government in many countries, and political leaders must take bold actions to remedy legal frameworks that are often based on pre-digital premises that are impediments to action. Politicians must not only enable change but personally drive the agenda.

2. *Digital doesn't care for our physical and analogue shortcomings.* Physical borders, people, processes, and economies have all shifted from the physical limitations we've had to contend with since the dawn of time. But most countries' legislation, policy development, citizen engagement approaches, and execution remain rooted in analogue thinking, limited by a rapidly archaic mental model.

3. *Nowadays, governments are impacted by megatrends and must continuously readjust.* A modern digital government must be deliverable on any platform (think Facebook or travel sites), on any device (think smartwatches or automobiles), and through ecosystems, due to the emergence of the Internet of Things. Governments must also quickly adjust to artificial intelligence impacts on their workforces and economies. These trends exist today, and governments around the world aren't reacting swiftly enough. Estonia is pivoting rapidly on these topics, and other countries must do so, too.

9

Government as a Platform: How Governments Can Lead in a Digital World

Alex Benay

WHY A GOVERNMENT PLATFORM?

The world has shifted. It is not in the process of shifting; it has already shifted. Platform economies have transformed many sectors: accommodations with Airbnb, transportation with Uber, retail with Alibaba and Amazon, media with Facebook and others, entertainment with Netflix and Spotify — the list goes on. The world is now dominated by platforms leveraging the interconnectivity of a planet that has seen a dramatic increase in mobility thanks to Internet of Things (IoT) devices, and soon, 5G. We are at the dawn of an exponential growth revolution.

A significant challenge in this metamorphosis is the uprooting of traditional economic, political, and social models. In his book *The Fourth Industrial Revolution*, Klaus Schwab addresses how every facet of our existence is impacted by digitization: economies, societal norms, biology, production, and numerous other fundamental aspects that shape our lives. "The digital revolution," he writes, "is creating radically new approaches to revolutionize the way in which individuals and institutions engage and collaborate."[1]

Nowadays, everything is digital — every new business is a digital business and all new research is data-science-driven in today's digital reality. For governments, however, the move to digital is much slower and more difficult, particularly for those who don't understand, both at the policy and operational levels, what this "digital" thing is all about. This is problematic because, as Schwab notes, "Ultimately, it is the ability of governments to adapt that will determine their survival. If they embrace a world of exponentially disruptive change, and if they subject their structures to the levels of transparency and efficiency that can help them maintain their competitive edge, they will endure."[2]

While the digital revolution is creating exponential growth all around us, what is both interesting and worrisome about this transformation is the relative absence of governments in the phenomenon. They continue to see themselves as different and immune to this change, even though its impact on citizens may be greater when it comes to governments. This makes the conservative stance by governments on digital change both dangerous and short-sighted. Around the world, governments provide many essential services such as research, social benefits, taxation management, and border control, but imagine how much more effective and accessible these services could be if they were delivered in a digital, platform-based manner. The reach and influence of public sectors around the world would be greatly amplified. Unfortunately, a long tradition of analogue thinking and a command-and-control culture stand in the way of governments accomplishing their full potential in a digital world. Historically, governments have had to develop linear-based approaches to service delivery and program design because the concept of mobility and interconnectivity for data simply didn't exist. Governments had a good reason to be linear and process-based.

However, in today's digital reality, governments can't afford to be linear. They must increasingly adopt exponential approaches to their work even if this isn't understood or accepted by traditional command-and-control systems of public service. By looking at Airbnb, Amazon, and others as examples to follow, governments can increase their productivity and achieve an exponential manner of thinking. This is where the concept of government as a platform (GaaP) emerges. But what does GaaP actually signify? It means changing the business model of government. It means dramatically increasing in real time the amount of data available to the general public. Today data is the currency and governments have more of it than any other sector. By releasing it to the public, governments can engage more freely and organically with stakeholders and citizens. Whether it is civil society groups who need data for urban planning projects, private-sector entrepreneurs who require government data to create an application, or researchers who want to collaborate on climate-change issues, pulling public servants out of the linear, analogue mindset dramatically increases real-time collaboration and results in more innovative output and greater growth.

This assumes that the right environment must be created for public-sector employees to collaborate with the "outside world," which directly challenges the traditional, hierarchical approach to absolutely everything in government operations. But it is a big shift that all governments need to get on board with. GaaP also assumes that many services could be "app-ified" by non-government entities, and such collaboration between government and its citizens should be seen as a healthy form of ecosystem creation. Realizing this opportunity, however, requires a massive culture change for many governments around the globe. It is easy to envision a service such as tax filing being conducted by several actors, or tax help forums and message boards being developed to enable massive online collaboration by the ecosystem in place of traditional mechanisms where public servants answer phones to provide services. But in order to achieve this, a fundamental rethink of the role of governments in a digital age is necessary.

Adopting such platform approaches can result in several benefits. An organization can truly leverage its people to do more and generate more meaningful work without necessarily increasing workloads. A platform approach to government can help develop entire industries. As governments release what they feel is useless data, companies can leverage it in more creative ways, as Ancestry. com or Flight Tracker have done. In addition, by switching to GaaP, public services worldwide can help raise the level of transparency and democratic engagement in government programs and services. This approach can also help make nations borderless and allow them to punch above their weights in a new, digital world where countries such as Estonia, Singapore, and Israel lead the way rather than traditional superpowers such as the United States and China. In fact, one can even claim that if governments don't adopt these tactics, they risk being left out of the new cryptocurrency regime or will fail to attract the kind of global talent they will need, meaning they could lose out on the potential to increase their GDPs or participate in new industries. Truly, this is about redefining the possible and the necessary in public administration writ large.

Simply uttering the word *change* in government can be difficult. Often governments are complex machineries that were designed pre-automobile. But change is possible. It is all around us if we choose to see it. What follows is such a change case. It was driven by necessity after a traumatic event: the full closure of more than 40 percent of the operations of a government crown corporation overnight. Consequently, the need for a fundamental change in its business model became imperative for the institution.

THE INGENIUM CASE

Ingenium is a Canadian crown corporation that houses the country's three national innovation museums: the Canada Agriculture and Food Museum, the Canada Science and Technology Museum, and the Canada Aviation and Space Museum. It has about 225 employees, ranging from historians and curators to information technology (IT) staff and exhibition interpreters, as well as overall management professionals. Ingenium has three distinct sites, all located in the Ottawa area. By global national museum standards, it is a very small institution in a medium-sized country. One of its biggest assets was the data and content it held within its walls: hundreds of thousands of artifacts, tens of thousands of archives, and much more content in various forms such as pictures or blueprints of planes, trains, and cars.

Even with all this content and data, however, Ingenium's approaches remained traditional and linear in design. It sold its pictures, it developed its historical assessments in isolation behind closed doors, and it didn't encourage staff to speak about their work publicly. It was like any other public institution around the world. As a result, its reach was historically below 10 million per year. By reach, I mean the government crown corporation touched someone either by social media, marketing, or other means — it meant people knew the institution existed in one way, shape, or form. Consequently, by not shifting its approaches to digital ones, Ingenium was, by all intent and purpose, a local entity with a national and international mandate. It is worth noting that during its 47 years of existence, the corporation had been hampered with limited funding and its buildings were crumbling. On a pro-rated basis, it attracted more visitors per year than any other national heritage institution but received some of the lowest levels of both operational and capital funding of any national museum, with the bulk of the funding going to the Canadian Museum of History and the Canadian Museum of Human Rights.

Nevertheless, in 2014–15, the institution managed to reach more than 11 million people. It did this in a variety of forms: over 3.5 million web page views and 3.5 million Canadians reached with its outreach and educational programs, as well as other initiatives such as 1.5 million reached with its "Let's Talk Energy" activities. From a program perspective, the corporation didn't see itself as a digital player or as having to embed digital management into its day-to-day activities. Consequently, its reach and impact remained low.

In September 2014, following the discovery of mould spores during routine renovations, the Canada Science and Technology Museum was closed indeterminately for repairs. Overnight, Ingenium's busiest museum was shuttered, and the corporation was forced to think of new ways to engage with the public. In 2015–16, a significant strategy entitled "Digital Citizenship at Ingenium" was created with the goal of completely changing the business model of the institution. After the creation of this strategy, a shift occurred and the organization began deploying digital-first mindsets to much of its work. The objective was to make the Ingenium network of private-sector partners, other museums, and academics into a powerful ecosystem. The goal was exponential growth of the establishment's reach versus linear program delivery.

As a result, Ingenium became the world's first public institution to truly work out in the open by adopting an "open-by-default" policy, which meant that all data and records were released within two hours of creation, whether they were drafts, final versions, or even raw data. The organization unleashed the power of the digital document-and-records system it had bought years ago and automated the release of everything. Management pressed for a top-down and bottom-up culture change. Initially, there was resistance. It isn't easy to work out in the open. But the concept requires a more democratic approach to cultural development and isn't a process in which government-funded institutions dictate what their patrons' heritage should be. So Ingenium sought opinions, input, and collaboration aggressively.

The outcome was the creation of the Open Heritage portal where documents would be released in real time upon creation.

Months later, tens of thousands of documents were circulated publicly: draft historical assessments, exhibition plans, and other raw contents were exposed for the world to see. Several naysayers had professed that there were too many "policy" restrictions to do this, that there would be too many complaints because government documents need to be final and complete before "publishing" on the Internet. All of these false risks were proven invalid. No complaints occurred and, in fact, what did happen was a dramatic boost in the museums' accessibility not only to the region and country but to the entire world. Being a digital government organization automatically means empowering people as global digital citizens. Journalists, historians in universities, and everyday history buffs began interacting with the institution in new ways, often discovering the museums for the first time.

Parallel to all this, Ingenium promoted aggressive utilization of social media by its employees, many of whom were already avid users in their personal lives. The aim was to remove traditional, outdated, command-and-control structures applied by government hierarchies. Now employees wouldn't need approval to tweet something about their work. To achieve this, management had to jettison an outdated mental premise that the Internet was only a medium to push content when, in fact, social media tools for quite some time had flourished due to the potential of two-way dialogue they created. In actuality, there was no better government sector than heritage and culture to encourage this kind of communication. The Ingenium experiment was a natural fit. Consequently, over time, a culture of online open dialogue was adopted by several staff members, engendering a culture in which people both inside and outside the organization could talk to one another. It was deemed acceptable to debate, engage, and talk to Ingenium's stakeholders, patrons, and the world at large. Respect remained key, since civil servants had to abide by a rigid set of rules, the same kind of respect guidelines applied in the analogue world.

The results were staggering. Total Ingenium reach exploded from more than 11 million to over 36 million in a single year. The organization's Let's Talk Energy Program saw its reach snowball from 1.5 million to 12 million in a year due to using YouTube, social media, short videos, and other media. Ingenium began seeing itself as a platform. Exhibitions were no longer the only means to connect with audiences. The employment of multiple media to deliver the institution's work began to shift the actual thinking for the programs and policies being developed. The concept of "going where others reside" emerged. Staff members involved in the National Outreach Program started attending comic-book and science-fiction fan conventions and festivals all over Canada to promote the science behind science fiction, while newer media such as video and mobile gaming and virtual reality were discussed in the halls of the museums as other means to attract the world on a far more significant level. The analogue and digital worlds were colliding and initiating, unbeknownst to the museums' staff, a new business model.

Another breakthrough in implementing a more "platform-based" approach to government materialized when a partnership was forged between SE3D Interactive, a small Toronto mobile gaming firm, and Ingenium. The objective was to share the costs of the production of new mobile games and divide the proceeds with this new private-sector partner. Ingenium knew nothing about the video-game sector or developing its own video games. In a traditional linear government system, originating its own games would have cost Ingenium hundreds of thousands of dollars and the risks would have been astronomical. But by releasing its content to the public, engaging with partners such as SE3D, and developing new, digital models, it was able to launch its first game for under CAD$25,000. The gaming company 3D-scanned First World War planes from the Canada Aviation and Space Museum and created a series of Great War flying games in which real air battles were re-enacted and the lives of pilots were honoured in an innovative

digital format on Android and Apple platforms for a fraction of
the cost of a traditional exhibition.

At the time when this book was being put together, three First
World War games had been produced. Each of the games were
downloaded in more than 180 countries in a matter of weeks, reaching hundreds of thousands of people globally, which is low for gaming standards but for the museums was astronomical. Ingenium also
began generating new revenues for itself as a result of partnering with
SE3D. In addition, the museums stopped selling their pictures and
data and instead promoted economic growth in the gaming industry,
revolutionizing their business model in the process. It is also worth
noting that the museums replaced one source of revenue with another — a more digitally based revenue model. The organization
started generating partnership revenue through video games, not
picture sales, unleashing the power of its content to drive innovation,
growth, and income not only for itself but for other parties, as well.

The latest iteration, *Ace Academy: Skies of Fury*, has been
downloaded more than a million times at a cost of production
well below CAD$100,000. To reach a million people through
an analogue model would have cost several millions of dollars in
production expenses. While certainly nothing replaces the "real
thing" — people seeing the museum's airplanes in person — what
Ingenium accomplished was to change its business model. It
didn't stop offering "the real thing," but complemented its artifacts with digital approaches, not merely digitization of existing
methods. The games reached millions of people who would never
come to Canada and permitted many to actually experience the
planes, battles, and history from anywhere in the world.

The result was continued growth of the museums' Internet
reach as it continued to alter its approaches to reflect new global
digital realities. In 2016–17, Ingenium reached over 38 million
individuals. This in turn led to record attendance at the Canada
Aviation and Space Museum. The pundits who claimed the

museums were losing their purpose and straying too far from their mandates were fearful of the unknown, proposing such hypotheses without truly understanding the digital world that surrounds all of us. This is an important lesson for governments everywhere: digital is a journey that requires attention to culture change first and foremost. Digital government is not about technology but about people. If it can happen in museums, places designed to preserve the past, it can and should happen throughout all government sub-sectors on the planet.

Ingenium's digital journey continues. Today the institution is investigating how it can produce 4K documentaries to engage an even broader audience. By releasing more content, connecting more online, talking to more partners, and creating an ecosystem of delivery as opposed to a linear delivery model, the crown corporation is able to find new spheres of influence and new partners. For example, the institution wanted to develop global documentaries for some time. As the institution investigated the potential of delivering them around the world, it discovered that producing a quality documentary would be costly, and Ingenium didn't have the expertise to deliver on this "new" medium.

By adopting a platform approach and developing an ecosystem instead of a traditional linear delivery method, the organization discovered that if it worked with a third-party provider, the third party could leverage federal tax credits for a majority of the total cost of production, drastically reducing the expense of producing documentaries to a few hundred thousand dollars. As it began investigating this possible partnership further, several management staff members were worried that the institution would develop a documentary that would "sit on a shelf" and not be distributed. After all, Ingenium wasn't in the business of distributing films. This was a valid concern. However, as it turned out, through the third-party partner, the executive team discovered there was an entire industry designed to distribute documentaries and television shows. In fact, that's how these companies got paid — by distributing content. Once again, the lessons learned are clear: engage third parties to create better value for money, expand the network of partners, utilize government content as currency to create a broader market, and leverage the expertise of others.

EXPANDING THE MODEL TO ALL OF GOVERNMENT

Critics will claim that Ingenium is a small crown corporation and therefore what it accomplished was easier. And they will argue that Estonia is a small country that was able to transform itself much easier. But these are hollow excuses. Let us instead focus on successful stories such as Estonia and Ingenium and figure out how to employ what they did on a larger scale.

What if governments today released their content by default? Imagine a world where all science is done in the open for everyone to contribute to. The power of our interconnected world could truly be brought to bear and the discoveries would no doubt be remarkable and happen at a much faster pace. Instead, scientific data and content is locked away in government vaults often never to be used, which is literally slowing down the progress of our species. Furthermore, imagine a world where heritage is truly co-developed by all parties, not driven by elites or the privileged. For too long we have told the stories of one class at the expense of another, rather than enabling people to develop their versions of the history or art presented in heritage institutions. Imagine a truly democratized heritage development process. The technology is available today to accomplish all of that. Not doing so is a choice.

What if governments, once this content is released, enable their staff to truly engage in online dialogue? Too many governments have been slow to enable their staff to talk on social media. Too many stories have circulated in which public servants had to seek multiple levels of approvals to tweet. This behaviour speaks to the culture of mistrust governments manifest concerning their employees, and it also speaks to the command-and-control structures in place in most public services around the planet. Such systems were designed pre-automobile and now exist in a world of mass collaboration. However, if governments are to be truly digital and participate in the exponential revolution, they must not only release content en masse in real time but also allow their staff to develop digital ties around them in order to operate more quickly. This can happen through the adoption of "social" government habits and practices in real-time dialogue.

What if governments, with their existing resource sets, focus those resources on changing the model to help drive an ecosystem instead of delivering in an analogue, linear fashion? We could create interoperability among systems and data to unleash economic development *and* services to citizens, releasing more content so others can leverage it and create opportunity. What if governments create digital ecosystems for all to participate and engage with in real time? Absolutely, cybersecurity is a real threat. But not engaging in the exponential revolution around it is also a key risk for governments everywhere.

Governments will never be experts in artificial intelligence, blockchain, cloud, and 5G, since those technologies are evolving much too rapidly. The only solution is to create a platform where content is available in real time, experts in government work with specialists in other sectors, and platforms and architectures actually enable collaboration en masse, again in real time.

What if governments evolve and enable the interconnected societies in which we live to amplify their outputs? To do this means more third-party service providers should be seen as acceptable alternatives. What if we "app-ify" government services and provide multiple options for citizens to engage with? What if passports are renewed through travel sites as an authenticated and trusted service? Releasing more content, enabling public servants to participate more freely and in real time, and redesigning how governments do business could lead to a very different, more IoT type of reality.

Imagine if the lessons learned from experiences such as Ingenium's are applied to larger institutions. What would be the impact on our economies? Could services be delivered through third parties and therefore be more in tune with the expectations of digital citizens? What if governments the world over become in tune with IoT and begin designing services and programs to be distributed on any platform on any device? This possibility will require a fundamental shift in business model from linear and ana-logue to exponential and digital. It will require governments to see themselves as platforms rather than continuing to view themselves, in an outdated fashion, as the backbones of their countries. It will require senior civil servants to understand that their national and regional governments, while crucial, are no longer the trusted and authoritative actors they once were. The world is interconnected, platform-based, and growing exponentially. If a few small museums can grasp this concept, governments around the globe absolutely can and should "get it" or risk becoming increasingly irrelevant.

*   *   *

ESSENTIAL TAKEAWAYS

In considering Ingenium's radical transformation, it is important to note that the crown corporation was undergoing a CAD$250 million facelift at the same time. It received CAD$80 million to relaunch one of its three museums and CAD$156 million to create a state-of-the-art storage facility, as well as millions of dollars in remedial infrastructure work across its three sites. The message is that digital doesn't need to occur at the expense of anything else. It does, however, require support by management as a "real thing." Management must allow its staff to engage online, must release more content in real time and provide the right tools to do so, and must first and foremost understand that digital is a new business model. It was never about digitizing museums and their collections; it was about working differently. The journey Ingenium has embarked on is a significant one and by no means over. Time will tell with the upcoming onboarding of new management if the changes brought on by the Digital Citizenship Program will stick or not. There are several lessons that can, even at this stage, be learned from Ingenium's experience:

1. *Content is the new currency and governments have this in spades.* They simply don't use it but rather keep their currency locked up in vaults never to be leveraged by the citizens they serve.
2. *Governments must learn to deliver through ecosystems in an exponential fashion, not a traditional analogue one.* This is necessary to keep up with changing skills and market demands brought on by the fourth industrial revolution and exponential growth.
3. *Ingenium's digital transformation was accomplished with no additional funding but rather with a reinvestment of existing resources.* It is the business model that must change. The process is about digital government, not digitizing government.
4. *Governments must create a platform for others to engage with.* Doing so can create significant economic growth.
5. *Culture change is the most important element when going digital and adopting digital government practices.*

The above points are only a few lessons learned and must be looked at in the context of broader government operations.

NOTES

1. Klaus Schwab, *The Fourth Industrial Revolution* (New York: Crown, 2016), 15.
2. Schwab, 69.

10

A Path to
Exponential Government

Salim Ismail

SALIM ISMAIL has been a great influence on my thinking and profession. His book *Exponential Organizations* has helped shape my management approaches and several digital strategies. He is a tremendously intelligent individual and his career is illustrious. For example, he is a board member of XPrize, a new challenge-based contest program designed to help solve complex global issues. Salim is also a founder of Singularity University in California, an institution focused on exponential concepts in an interconnected society. For those reasons, I am delighted to have him write about what "exponential government" should look like and how governments can become platforms for others to create on, to grow from, and to essentially cease being linear and start being exponential.

– Alex Benay

This chapter has an obviously grand and aspirational title, but it matches the rationale for this book. The previous chapters articulate eloquently some of the structural issues governments face in an increasingly digital and accelerating world. More important, many chapters describe well the target to aim for, be it fifth-generation (5G) communications or digital democracy. The sections that follow in this chapter summarize the fundamental shift occurring as well as a prescriptive path on how to achieve exponential government.

EXPONENTIAL GOVERNMENT — AN OXYMORON

A highly illustrative example of the challenges facing governments comes from my previous role at Singularity University (SU), an institution that brings together the leading minds in the fastest-moving technologies. Perhaps the most unique feature of SU is that the curriculum is updated in near real time because the underlying technologies move so rapidly. For example, someone doing a master's in neuroscience today will find the knowledge acquired out of date by the time the degree is finished. The changes in many subject areas (think blockchain) progress far quicker than the ability to synthesize and teach them. So to be relevant, SU was forced to create a novel method of near-real-time curriculum development. But here's the problem: to become an accredited, government-sanctioned educational institution, curricula must be fixed and not changed! So despite having the leading scientists, researchers, and thought leaders globally coming to teach, SU can't be government-approved.

Today there are a multitude of such structural issues facing governments in every capacity. The previous chapters explain very well the need for change and provide compelling views of what's required. What is presented here is the path from A to B. How do we shift from sub-linear policy-making and enforcement to an enabling, testing, fast-moving, accelerating environment? But first, let's summarize the structural issues we're facing.

LINEAR TO EXPONENTIAL EQUALS SCARCITY TO ABUNDANCE

The doubling patterns we've witnessed in computing for the past 60 years are now being seen in about a dozen technologies, some of which Iain Klugman identifies in Chapter 2. But add to his list biotech, solar energy, autonomous cars, and drones and we see the larger picture. Importantly, while each technology doubles every year or so, where they intersect adds a whole other multiplier to the equation. The lidar unit (laser radar) used by autonomous cars cost US$75,000 in 2012. In 2017 that same unit cost US$50. Why? Because there are multiple embedded technologies, each of which is accelerating, so the aggregate compound effect of these intersections is enormous.

More important, governments that used to slow down the advancement of technology with policy are unable to do so today. For example, President George W. Bush attempted to block stem-cell research in the United States for ideological reasons by stopping federal funding. As a result, many U.S. researchers moved to Canada, China, or Australia, and stem-cell research continued at the same pace while the United States fell from number one to number eight globally in this area of science. In a more dramatic example, the European Central Bank said in November 2017 that it would be unable to regulate cryptocurrencies. This was a staggering, if accurate, admission. Furthermore, due to the pace of change, by the time a particular technology is regulated today, its evolution is several generations ahead.

A second fundamental characteristic of technology is that it transforms a domain from scarcity to abundance. In their book *Abundance: The Future Is Better Than You Think*, Peter Diamandis and Steven Kotler tell the story of aluminum, which 200 years ago was the most expensive precious metal in the world because electrolysis — which separates aluminum from bauxite — hadn't been invented yet. These days we throw away tinfoil without even noticing.

Today, given that we have a dozen technologies accelerating, it means that entire industries shift from scarcity to abundance in short order. Perhaps no better example exists than solar energy, which is doubling its price performance every 22 months and has been doing so steadily for 40 years. At that pace we'll be able to deliver 100 percent of global energy supply using solar in 14 years. Even cargo ships now employ battery technology. So energy, which has been scarce for the entire history of humanity, is about to become abundant. Just in 2017 we saw a 30 percent drop in solar systems, and Chile already generates so much solar power that it gives it to neighbours for free. The poorest countries in the world are the sunniest ones on the planet, so the geopolitical impact will be profound. Globally, I predict the CAD$6 trillion energy industry will shrink to CAD$2 trillion in the next 20 to 30 years.

For Canada, about 40 percent of total exports are oil. So doing nothing or moving slowly is clearly not an option. In fact, the Canadian government needs to adapt fast and dramatically. The luxury of slow, incremental changes over decades is gone.

THE IMMUNE SYSTEM

The fundamental difficulty in implementing disruptive change is that when attempted, the organization or institution being affected attacks any new idea with its "immune system" in the same way bacteria invading a biological entity are targeted and destroyed by the entity's immune system. And it's a function of size — the bigger or older the institution, the more established its immune system. Any change management consultant can show the scars on his or her back from the multiple arrows from these types of assaults, resulting in a very short tenure for anyone in that function.

While disruption is difficult in the private sector, where vested interests will stop new ideas (think Kodak and the digital camera), it's much tougher in the public domain, where existing policy often *is* the immune system. Efforts to update education systems will invite aggressive pushback from teachers' unions. We see the same with taxis and Uber/Lyft.

Mary Francoli shows in Chapter 1 the struggle of representative democracy to keep pace with an accelerating world, but this extends to all institutions. Take, for example, marriage, which first surfaced thousands of years ago. In the Iron Age, the average human lifespan was about 25 years. So typically people got married, had kids, and then died! Marriage — and indeed evolution — was never designed for lifespans of more than 80 years. And what will happen when we double human lifespan in the next decade or so?

We also see this issue in the orthodox populations of religions today. Be they Hindu nationalists, Christian evangelicals, Muslim radicals, or Jewish settlers, pockets of people in every religion clamour for a "return to values" because they can't handle the pace of change. Today this is the primary cause of human conflict and war, which most would agree is a very inefficient way to progress humanity.

SIZE MATTERS, BUT IN THE WRONG WAY

In the past few centuries, it was an enormous advantage to be a big country. Large nations had lots of natural resources, critical mass in multiple industries, and so on. But today big countries are under enormous stress, since the complexity of managing them has exploded. Russia, China, the United States, Brazil, and India are all heaving masses of chaos. Their size, which was advantageous in past eras, is a liability today. We see most governance and actionable progress occurring in small countries, even at the municipal level. Consider that cities such as Tokyo, São Paulo, or Mexico City are more populous and more complex than any country was a century ago. In Canada, note how the provinces, which control budgets, hold their cities hostage to progress.

LEADERSHIP TO LEADERS-HIP

The final challenge in this current age is leadership. In Chapter 5, Hillary Hartley summarizes this need quite well. Consider that our leadership structures and training were designed for a scarcity-based, linear, incremental, material world and we're entering Abundance Central Station. Even in the private sector, there are no master of business administration programs in the world that can instruct us how to create an Uber or Airbnb — they all teach how to build 20th-century businesses. So to summarize the problem:

1. There are a dozen technologies accelerating and acting as forcing functions.
2. Our governments (and, indeed, all our institutions) weren't designed for this era.
3. Real change occurs at local and city levels, not at the national level.
4. Our existing leadership training is inadequate.
5. Attempting disruptive innovation provokes attacks from the immune system.

So we need to shift from operating in a predictable world, where the scaling of efficiencies was the dominant strategy, to a world in which adaptability and disruption represent higher-order competitive advantages. Clearly, this is a thorny problem, so let's look at how to address it with three simultaneous parts of a solution.

1. Broad, Aspirational Policy, Locally Implemented

In the future, we must set broad policy at the federal level and allow implementation at the local level. There is a precedent for this in the military called "commander's intent." Broad objectives are set, and local leaders have the authority to achieve targets as they see fit because they have better and more current situational awareness. The United Nations shows such leadership in its sustainable development goals (SDGs), which are highly aspirational but allow local execution.

In the private sector, this is known as the massive transformative purpose (MTP). Examples are Google ("Organize the world's information") or Uber ("Everybody's private driver"). As an instance of implementation, leading human performance management systems use objectives and key results (OKRs) in which a broad objective is set, but the team leader and even individuals have a broad mandate about how to accomplish success. This approach is employed fully in fast-growing companies such as Google, Facebook, LinkedIn, and Twitter.

2. Disruptive Innovation at the Edge

The second piece is to find disruptive, new solutions that leverage the potential of new and accelerating technologies. It's clear that this can't be done inside existing institutions for two reasons. First, if attempted, the immune system will attack. And second, legacy thinking inside existing structures will only yield incremental results. Disruptive innovation always comes from outside that industry or sector (think Tesla or bitcoin).

The easiest way for governments to embrace such change is to aggressively sponsor entrepreneurship ecosystems, something being done very well across Canada.

3. Solve the Immune System

The third piece is to solve the immune system problem. It's here where we drill into some detail and examples, since this is a relatively new problem space that hasn't been addressed systematically in the past.

ExO Sprint

In 2015 a company I founded called ExO Works created a 10-week process called ExO Sprint, which aims to solve the immune system problem in corporations. We piloted it with Procter & Gamble, have since run it a dozen times in blue-chip corporations, and find we can move the management, leadership, and culture three years ahead in that 10-week period. The process works as follows:

- Run a half-day workshop with all senior management of the organization. Showcase new technologies, threats, and opportunities — very much shock and awe — to demonstrate that disruptive threats are on the horizon and something must be done. This creates a burning platform for change.
- Gather 20 to 25 young leaders/future lieutenants of the business. They do the work over the 10 weeks and are divided into two streams. One stream looks at disruptive new ideas in adjacent industries that could grow the business by 10 times or more. The second stream examines the existing organization and chooses mechanisms to improve the status quo. At the end of the 10 weeks, they present their ideas, and senior management funds those they believe are worth it.

In analyzing the reasons for success, we've found that the opening workshop acts like an immune-suppressant drug. When doctors perform kidney transplants, they administer immune-suppressant drugs so that new kidneys have time to bed down. We've found similar results in which the normal attack of the status quo is suppressed and new ideas have time to get footholds.

The best quote we've received for this process comes from the CEO of Rassini, a global company of 11,000 employees making car parts for Tesla and Mercedes, who said: "You took my white blood cells (people) that attack new ideas and turned them into red blood cells providing oxygen." Also, by having future leaders create new ideas (with coaching support), they champion and own them into the future, increasing chances of adoption. Where in the past disruptive ideas get funded 10 to 15 percent of the time, ExO Sprint results in 97 percent of new ideas being fully funded.

Fastrack Sprint

After seeing success in the private sector, the question then arose of how to implement a similar process in the public sector where existing policy is often the immune system and thus much more difficult to address. In 2016 we formed a non-profit organization called Fastrack Institute, focused on this effort. The objective is to implement aggressive new ideas into cities while solving the legacy policy issues (i.e., the immune system). One of the co-founders is Rodrigo Arboleda, who was the CEO of One Laptop Per Child and has considerable experience implementing technology in difficult regulatory environments.

We adapted ExO Sprint for the public sector and created what we call Fastrack Sprint for cities and regions. The objective is deliberately aggressive: to drop the cost of an existing domain (e.g., transportation, education, et cetera) by 10 times. Teams are formed that take a problem space (e.g., transportation, education, health care) through four phases:

1. Technology — Examine Breakthroughs That Drive the Future

- Work with maker spaces, bio-hacking and fabrication laboratories.
- Examine accelerating technologies — sensors, 3D printing, robotics, artificial intelligence, synthetic biology, drones, et cetera, that apply to the problem space.
- Look at smart cities technologies to advance services, infrastructure monitoring, traffic flows, new transportation systems, health systems, et cetera.

- Green building construction and renovation, low-carbon energy sources, smart grid, energy-efficiency measures.

2. Design — Picture, Imagine, Design, and Describe Technology Solutions

- Artists, science-fiction writers, designers, and media experts envision and paint a future with technologies in mind.
- Design experts use human-centred techniques to integrate the vision into possible products and services.
- Urbanists with deep knowledge of cities, from economic structures to city planning, design what implementation looks like.
- Industry experts convene to work with this group to run scenario planning exercises about the future.
- Media/narrative experts weave a story around that picture of the future.

3. Entrepreneurial Layer — Ensure Economic Sustainability

- Combine design and technology to solve major problems with a business model.
- Develop funding mechanisms to help entrepreneurs raise money.
- Ensure sustainability of potential solutions (via business or taxes).
- An accelerator and seed funding are part of this layer.

4. Social Layer — Implementation into Society

- Sociologists, anthropologists, regulatory experts, and legal thinkers consider how solutions can be integrated into society.

- Explore public/private partnerships for deployment of ideas.
- Experiments and trials with the projects, ideas, technologies, and companies come from other layers.
- Community leaders represent different social groups, classes, and interests.
- Extensive gathering of data with a feedback loop aids iterative improvement.

Incentive prizes are used in each phase to encourage numerous and diverse pictures of the future. As with ExO Sprint, teams are formed that take a problem space through these layers and develop solutions. We intend to open-source the process, and it's envisioned that different regions/cities will create their own replicas, with best practices shared among all. What follows are two examples.

Our first city was Medellín, the capital of the department of Antioquia in the Aburrá Valley. The city and its citizens have shown determination and resilience in the face of complex challenges. Today Medellín is recognized globally as one of the most innovative cities in Latin America as well as in the world. Its successes attracting companies, technology, venture capital, and young talent have been impressive. The city has undergone an incredible transformation in the past 20 years, shifting from a fear-filled and uncertain past to a place that has regained hope for a positive future. Medellín has launched multiple social inclusion projects such as metro cables connecting rapid-transit lines to the top of mountains, electric staircases running through entire neighbourhoods to shorten commutes, and public areas enhancing tightly knit communities.

We raised funding from local foundations interested in solving problems in the city and created a partnership with Ruta-N, the innovation arm of the city. Then we ran Fastrack Sprint for the following problems: mobility, financial inclusion, health care, and air quality. The results have been gratifying. Several companies were created and are being accelerated by local teams that have solicited investment for these start-ups.

After witnessing success in Medellín, we focused on a major U.S. metropolis. Miami is home to more than 1,400 multinational corporations and acts as a significant centre and leader in commerce, culture, finance, tourism, media, international trade, and arts and entertainment. Miami has begun to absorb key entrepreneurial, technological, and creative trends, becoming a hub for start-ups and innovation that attracts core investors and businesses to generate a positive impact on a global scale. Miami-Dade County, home to more than three million residents, continues to experience spectacular growth and development spanning over the past two decades. The growth has come with a corresponding traffic problem, and Miami is considered one of the top 10 most congested cities in the world.

We ran a Fastrack Sprint in Miami, forming teams to explore the problem. A Regional Advisory Board was developed with members of local government, the county Department of Transport, academics, entrepreneurs, and foundations. The process took 16 weeks, finishing in December 2017, with the following four components of a solution:

- **Immediate congestion alleviation (within a year).** A ride-sharing app and data analytics platform was proposed with heavy social engineering and points to reward ride-sharing in busy corridors. Tolls would be lessened outside rush hours to smooth out the traffic hump. Discussions with schools were suggested to change start times. Video feeds from existing closed-circuit television (CCTV) cameras would be used with image processing and video recognition to count cars and set up an environment to create what-if scenarios.
- **Test last-mile/first-mile technology solutions.** Letters of intent were signed with several self-contained communities in Miami (Coral Gables, Doral, Miami Beach) and with

technology vendors (electric bicycles, small electric cars, and autonomous cars). These will be experimented with and where success is seen will be merged into the broader city.

- **High-throughput corridors.** Instead of a decade-long request for proposal (RFP) process, a switch to requests for information (RFIs) will be invited from major vendors to suggest solutions for rethinking chief transportation corridors. Current projections are that new solutions (e.g., magnetic-levitation trains) can be implemented at a tenth of the cost of traditional rail.

- **Policy changes.** A multi-sector policy team was proposed to route around state and federal regulatory bodies (e.g., federal interstate rules) and gain exceptions.

Initial response from the mayor's office and key stakeholders has been completely positive, the solution is set, and funding has been made available for immediate development/deployment. This challenge provides a golden opportunity to shape Miami's future as well as help the city to become a leader in mobility by solving mass-transit issues.

After five such Sprints in cities with extremely complex problems, we're excited that there exists today a systemic approach to address major problems with new technologies and radically transform them.

AN EXPONENTIAL SOCIETY

A fundamental shift is happening globally that can't be ignored. Technology is creating extreme stress in society and either we surf the tsunami or get crushed by it. Disrupt or be disrupted — there's no middle ground anymore. It's exciting, however, that we might have found a systematic approach to solving the immune system problems of both the private and public sectors.

Most important, though, is that I keep hearing how Canada has an amazing opportunity to lead the world, not as an opportunity but as an obligation. Canadians are incredibly lucky to live in the country they have and it's our duty to use that privilege to lead the digital transformation.

I travel globally to meet heads of state and corporate leaders, and there are clearly two stark futures in front of us: either we find ourselves in a *Star Trek* or a *Mad Max* future. Today, sadly, we seem headed toward the latter. I believe our duty is to leverage Canada to tilt the world into a future of abundance.

*   *   *

ESSENTIAL TAKEAWAYS

Salim Ismail outlines the opportunities and challenges for governments to reach exponential thinking. We live in an age of growth that's happening at breakneck speed due to the digital realities addressed in this book. An interconnected society, 5G, the rise of mobile devices, and other technological advancements are putting all sectors, whether they see it or not, on a path where traditional mindsets and controls no longer apply. We must, as governments, alter our thinking, approaches, and methods. Here are some key takeaways:

1. *The old truths are no longer relevant.* Big scale, which used to be an advantage, is now a hindrance. Once upon a time, large governments were seen as sources of stability and predictability for countries. Now the same big bureaucracies often impede the ability of governments to react fast, innovate, or even partake in the exponential digital economy they often regulate. It is not about

digitizing analogue government but rather completely re-imagining it in a digital age.

2. *Cultural change is the key.* No amount of technology will help governments if they don't transform their mindsets to embrace the exponential growth we see everywhere, embrace a digital-first approach, or simply change a process. In a digital and exponential world, standing still actually means regressing. To adjust to this new reality, new leadership models are required, new teachings must emerge, and new ways of thinking must become the norms, as opposed to anomalies.

3. *True government digital change can only happen by removing access barriers to government.* Real change can't happen from within, or at least solely from within. Governments must first want to re-imagine themselves and then remove all the excuses that stop them from interacting with the outside world. Governments can't speak to the world through procurement exercises, "consultations," or programs. We must talk, engage, debate, discuss and, most important, find solutions with experts who are almost always outside government when it comes to digital transformation.

CONCLUSION

WHAT WE'VE COVERED

The chapters in this book lay out a basic approach to government digital. As we've seen, governments have suffered from a lack of public trust for decades now, a phenomenon only compounded by the new digital reality we all live in. Mary Francoli, a prominent expert in open government and an active participant in the Open Government Partnership network, makes this argument quite clearly in Chapter 1.

In Chapter 2, Iain Klugman illustrates the impacts of the arrival of the fourth industrial revolution — new medical breakthroughs and greater wealth produced by fewer people, as well as the need for governments and countries to make a complete and abrupt course correction in multiple sectors of activity. This leaves public services around the world attempting to determine the best course of action. The speed of change is one of the greatest challenges to modern public governance in this century.

Ray Sharma and Amir Bashir show in Chapter 3 how fifth-generation (5G) networks will usher in an age of unprecedented growth that will allow us to accomplish great things as a global, interconnected society as long as cities, counties, provinces, states, and nations become fully digital. Governments must continuously invest in technology and deliver new real-time services in areas crucially important to their citizens, and that means keeping on top of all of the latest advancements and implementing them effectively into health care, education, transportation, the environment, finance, and so much more.

In the education sector, as presented in Chapter 4, John Baker suggests how learning is now a lifelong mobile endeavour that must be supported by not only the right tools but also the correct alignment within this rapidly changing world. Coding, analytical skills, and other 21st-century approaches must replace the current curricula in many instances, and immediately, if governments wish to pursue modern, digital economies and societies writ large.

In Chapter 5, Hillary Hartley speaks of the importance of governments to "work in the open" as public agents of digital change, a necessity that's imperative. For years public servants were told to be quiet and merely assist in governing nations. In a digital, exponential, and interconnected society, government employees must increasingly leave the comfort of their cubicles to engage with stakeholders to deliver new digital programs rather than digitize old, outdated, analogue processes. This can only happen if public servants use modern tools and transparency to develop policies, solutions, and services with others. If Sharma and Bashir define the interconnected society, Hartley addresses the need to "plug into" this emerging global trend if governments are to remain relevant.

Jennifer Urbanski, in Chapter 6, shows how online engagement by public-sector employees at all levels can help promote working in the open, create new dialogues, and possibly build ecosystems capable of delivering public-sector innovation through the democratic act of connecting with stakeholders. A mere decade ago we didn't possess the tools available today. Activists have understood the importance of these social tools, launching massive global movements such as the Arab Spring and Occupy Wall Street, yet often government engagement is tokenistic with messaging prebaked so that we rarely accomplish true two-way dialogues. Governments must stop using social media solely as an information dissemination tool.

In Chapter 7, Olivia Neal makes the case for a new type of government service design approach whereby digital first becomes a core tenet. With the purchase by Amazon of Whole Foods and the initiation of Amazon Key to deliver goods inside our homes or vehicles, the very concept of service delivery has changed and should be transformed for governments, as well. As Scott Brison, president of the Treasury Board of Canada, often claims, "We cannot be a Blockbuster government serving a Netflix citizenry."

In Chapter 8, Siim Sikkut relates how Estonia, a small European nation, conscientiously decided to make itself digital rather than merely digitizing the country. The state became virtual by providing an e-Residency program that aggressively adopted business-creation incentives so that entrepreneurs never have to set foot in the country, while at the same time investigating how cryptocurrency could replace the national, analogue currency system. Estonia reveals what the evolution of nation-states around the world will look like in a digital universe.

Once governments understand that they must become *the* platform to engage stakeholders and achieve effective delivery, amazing things will happen. In Chapter 9, I showcase my attempt to turn Ingenium, a cultural institution, into a platform, and the numbers speak for themselves. The reach expanded exponentially, partnerships delivered real value, new stakeholders were engaged, and transparency was increased dramatically, all occurring without any new budget money incurred. It was a question of shifting how we delivered services while not adding to the analogue way of doing things — an important lesson about changing the model of government.

Finally, in Chapter 10, Salim Ismail offers a blueprint for how governments must become exponential and holds no punches in discussing how governments around the world can't hide behind big scale in a digital age. In fact, being small, nimble, and agile is a greater advantage than being large and complex, whether in business or government.

SO NOW WHAT?

We've heard from practitioners leading the way digitally in their respective fields. Experts from Canada, the United Kingdom, the United States, and Estonia have spoken on various digital government topics, all as important as the next. The concerning fact is that by choosing the subject matter for this book, many topics — cybersecurity, for instance — were left out or touched on only briefly. We also could have chronicled the amazing work being done in countries such as New Zealand, Israel, and South Korea, but again space constraints made that infeasible. Perhaps a second book could expand the dialogue even further. But since digital literally affects everything governments do now on a daily basis, such as regulations, taxes, services, and policies, determining where to stop the dialogue is likely futile. There are, however, eight core principles that do emerge:

1. **Digital is everything a government does.** Countries must change their lenses because digital and technology are no longer back-office functions. Digital must be at the beginning of every discussion when developing new policies, programs, or services. For this to happen, public services around the world must bring on board new digital work skills and knowledge. This skill set is often lacking and must be developed, recruited, and nurtured. It's no longer about digitizing content; it's about building a completely digital government, which implies doing things drastically differently.

2. **Linear approaches to problem-solving must be replaced with exponential engagements**. When governments identify problems, they spend years defining requirements, then years more mired in various procurement exercises. If they're lucky, they deliver on their commitments, but more often than not, the projects fail. This is linearity at its worst. Luckily, there is a different way to work that must be explored in the digital age. Governments have to join open-innovation ecosystems and identify their problems, not their needs, when engaging with the world around them. The pace of change is simply too great to waste years in typical linear approaches. Governments should release as much content as possible to enable innovation by third parties and not allow this data to languish behind firewalls. Then they must empower their employees to solve problems in the open and link with all sectors to develop common global solutions to navigate the fourth industrial revolution depicted by Iain Klugman and harness the interconnected 5G society described by Ray Sharma and Amir Bashir.

3. **The legislative and policy landscape must change.** Privacy is a key topic that isn't explored fully in this book, nor is access to information, data-sharing across organizations, and enabling citizens to own their personal data and control how it's shared with governments for gains in efficiency of service delivery. These are important subjects that must be studied through a digital lens. For example, if governments become increasingly open by default, essentially releasing data upon creation, are freedom of information acts required in their current iteration, or is a different type of legislation needed? The current legislation that applies in an analogue world can't be carried over as it is into the new digital reality. The secret recipe for countries to achieve "Estonia-like" success is to be willing to quickly adjust their legislative frameworks to operate in the digital age.

4. **Governments must operate in the open.** When public servants are empowered to work, create, discuss, and solve problems with others, they actually perform the role of civil servants in the 21st century. The civil service of the 20th century was different because the world and its technologies were different. Public-service employees couldn't mass-communicate the way they can today, nor could people work from anywhere, anytime, as they can with today's tele-communication technologies. The world has shrunk and too often governments aren't connecting their workforces to the world around them, essentially missing out on enormous pro-ductivity possibilities. Governments are no longer the single version of a truth, or even the holders of a trusted inform-ation source. Digital governments must engage with other sectors every single day, but not through outdated consulta-tion processes, because the planet is changing exponentially. Expertise one day is outdated by the next, and if public-sector employees are to operate in a digital world, they must accept the fact that all other sectors are likely ahead of gov-ernments moving forward and that true digital policies, service delivery, or other programs can only be achieved by governments partnering with the world at large.

5. **Government employees will never again be the experts in anything in today's digital reality.** We've reached the age of the fourth industrial revolution in which exponential growth in health care, energy, technology, and economies are moving too rapidly. Currently, with mostly linear approaches, govern-ment organizations simply can't keep up with these changes. Instead, public servants must see themselves as enablers, con-nectors, and networkers who can identify local, national, and international issues and quickly marshal the connected brain power of stakeholders to resolve them. This is a different role than most governments are accustomed to, but it's the new

reality we all face in light of the exponential growth curve before us, one that, ironically, is often paid for by public funds.

6. **The public service must increasingly adopt open-source software.** Governments around the world can now collaborate as never before. Paying taxes, registering a birth or death, launching a business, and countless other services are often not very different from one another in democratic countries. This means by increasingly adopting open-source technologies, national, regional, and local governments can not only share solutions but co-develop the actual services with their key stakeholders. Open-source technologies are a new normal in any enterprise architecture for any government at any level, but too often chief information officers haven't made the shift and still rely too much on proprietary software, which stifles mass development of solutions.

7. **In a digital world, the public-sector human resources model must change.** The gig economy is here to stay. Workforces are shifting and often governments still see their employees as 20- to 40-year assets, depending on the length of a career for a typical public-sector worker in any given country. We must enable rapid in-and-out of public servants to leverage new and emerging skills, technologies, and approaches. The reality is that governments often take too long to recruit. If it takes a year to hire and six months to train, the opportunity is often lost. To pivot more quickly, governments must dramatically alter the human resources model.

8. **Governments must adopt a digital-first mindset for their service delivery.** Services in our personal lives are, for the most part, digital or digitally enabled. Yes, there will always be a need for analogue service delivery, but even these mechanisms are digitally enabled with logistics management, tracking, or other apparatus. With the increasing pace of change and growing citizen expectations, services can be delivered on any device, on

any platform, and within a broader ecosystem. In an age when individuals can get their services delivered and managed from their mobile devices, whether it's food delivery, transportation, or banking, it's presumptuous for governments to assume that citizens must come to them for services. Government services must become ubiquitous as quickly as possible.

The above eight realities should become core strategic elements of any government or public service attempting to embark on the digital journey. The road is already complex, and it's more complicated in government. I can personally attest to that, having worked in both the public and private sectors in some digital capacity or another. Government change is simply harder to achieve. In many ways, we should be happy that's the case. Governments must find the right balance between being on the leading edge and ensuring that their countries continue to operate. However, the biggest challenge for public sectors around the world is that if they don't change quickly in many aspects and continue doing things the way they always have, the result will be catastrophic failure sooner rather than later.

In essence, what exactly do governments risk by going digital — projects failing? That already happens today. In the grey zone between digitizing analogue and going for truly digital government, public-sector projects fail and will fail. Private-sector projects fail, too, but the difference is that the scrutiny is greater in government because of the social and economic impacts that come with the job. The overall point is that failure is already here. Therefore, the risk of making the digital journey must be discussed in context.

I believe that the risk of government inaction is far greater than the danger of failing to make the shift to the digital age. How will we communicate with our banks, which are all aggressively investing in artificial intelligence as we speak, if we haven't yet begun adopting these technologies in our public environments? How will nations compete with the likes of Estonia for talent or resources when that small Baltic state has decided to remove physical impediments and barriers to stimulate its economy, instantly tapping into the global interconnected society? How will national skills and education systems that are often archaic not drag their countries' GDPs down with their obsolete curricula in the face of the digital revolution? Digital isn't a back-office "thing"; it's at the very heart of government in every way, shape, or form. We must treat it with the utmost respect and as a top priority across all levels of government or risk suffering dire consequences in the very near future.

— Alex Benay

ACKNOWLEDGEMENTS

I would like to thank all the amazing contributors who donated their precious time to this book. They are all incredible leaders in their respective fields and are clear representations of how diversity as well as inclusion can lead to better outcomes for governments and ultimately for citizens. Furthermore, this project wouldn't have seen the light of day without the support of key people: Ciara Cronin, Tasha Ridell, and Katie Wyslocky, who each took turns helping to coordinate numerous elements of this project. *Government Digital* wouldn't have been possible without their support.

I would also like to thank the Ingenium Corporation, which manages Canada's three national science and technology museums, as well as the Ingenium Foundation, which will be using the funds raised from this book to create much-needed programming for women in technology in the hope of increasing gender parity in the technology sector. This is a global problem, and certainly a core issue within the science and technology sector in Canada, that we all need to concern ourselves with. I hope this book can lead to better programming for our young women to encourage them to take up the cause and make the world a better place.

Lastly, I would like to thank my family. Choosing a career in the public service isn't as easy and plush as many might think. I've had the privilege of working in both the public and private sectors, and let me tell you, the public sector is harder in every way possible. Often, yes, that can be due to our own fault and outdated ways; nevertheless, it is harder.

For someone to have a successful career in the public sector, you need a tremendous support network. I'm fortunate to have a wife who backs all my crazy schemes and two amazing children who will grow up to leave their marks on the world. Gen, Cloee, and Gabe, thank you for supporting me throughout all these years, permitting me to focus on giving my very best to the calling that is the public service.

— Alex Benay

LIST OF ABBREVIATIONS

AI: artificial intelligence
API: application programming interface
AR: augmented reality
AV: autonomous vehicle
CBE: competency-based education
COBOL: Common Business-Oriented Language
CODE: Canadian Open Data Experience
COIN: Contract Intelligence
DFAIT: Department of Foreign Affairs and International Trade
  (Canada)
D2L: Desire 2 Learn
FCC: Federal Communications Commission (U.S.)
FOSS: free open-source software
GaaP: government as a platform
GCHQ: Government Communications Headquarters (U.K.)
GDP: gross domestic product
GDS: Government Digital Service (U.K.)
GTA: Greater Toronto Area
ICT: Information and Communications Technology

IoT: Internet of Things
IT: Information Technology
ITU: International Telecommunications Union
L&D: learning and development
MOOC: massive open online course
M2M: machine-to-machine
MTP: massive transformative purpose
ODS: Ontario Digital Service
OECD: Organisation for Economic Co-operation and Development
OGP: Open Government Partnership
OKR: objectives and key results
OPS: Ontario Public Service
OTT messaging: over-the-top messaging
PISA: Programme for International Student Assessment
RFI: request for information
RFP: request for proposal
SDG: sustainable development goal
SME: small- and medium-size enterprise
UX: user experience
VR: virtual reality

Canadian Failures
Stories of Building Toward Success
Alex Benay

The Hill Times: Best Books of 2017

Successful Canadians write about failure, and
how it got them where they are today.

What does it mean to fail? To some of the most successful Canadians, it was a rite of passage, a stepping stone to greater things,
or even a brilliant source of inspiration. Olympic golds, successful
businesses, pioneering medical advances — all came about after a
series of missteps and countless attempts.

Canadian Failures gathers ten experts from the private, public,
and not-for-profit sectors and academia, all of whom have grappled with failures and success throughout their lives. Their powerful argument: that Canada, and Canadians, must be willing to
learn from failure if we hope to succeed.

OF RELATED INTEREST

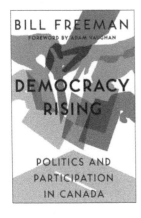

Democracy Rising
Politics and Participation in Canada
Bill Freeman
Foreword by Adam Vaughan

The Hill Times: Best Books of 2017

What are the limits of Canadian democracy and how are they being expanded by a revolution in participatory democracy?

The Brexit vote in Britain and the election of Donald Trump as president of the United States illustrate that our system of representative democracy is in deep trouble.

There are signs of political alienation everywhere. Most believe that government is run by a few big interests. Wealthy corporations receive grants and beneficial regulations. The incomes of middle and lower earners have remained stagnant or decreased.

The way to change this imbalance is by strengthening our democracy and encouraging participation in the political process. A powerful grassroots movement of participatory democracy is emerging. Freeman's message is that democracy is rising in this country, but we must organize to redress the dominance of business interests and fulfill the promise of government by the people.

Throw Your Stuff Off the Plane
Achieving Accountability in Business and Life
Art Horn

A guide to making the leap from imposed accountability to personal commitment for both individuals and organizations.

Accountability — we all want the people around us to be responsible, reveal genuine commitment, keep their word, and stay away from blaming others. But organizational systems that aim to institutionalize accountability don't quite go all the way. People are people. They have their own wants and needs, their own psychological tangles, and they often don't particularly want to be held accountable, let alone confront others who have let them down.

Throw Your Stuff Off the Plane is here to help. It reveals the missing ingredient organizations usually overlook: personal responsibility. It's an approach to self-improvement for each reader, centring on untangling the conflicting thoughts that block personal responsibility. And it's a guide for every leader who wants to go all the way.

Book Credits

Acquiring Editor: Beth Bruder
Project Editors: Kathryn Lane and Elena Radic
Editor: Michael Carroll
Proofreader: Karri Yano

Designer: Laura Boyle

Publicist: Kendra Martin/Michelle Melski

Dundurn

Publisher: J. Kirk Howard
Vice-President: Carl A. Brand
Editorial Director: Kathryn Lane
Artistic Director: Laura Boyle
Director of Sales and Marketing: Synora Van Drine
Publicity Manager: Michelle Melski

Editorial: Allison Hirst, Dominic Farrell, Jenny McWha, Rachel Spence, Elena Radic
Marketing and Publicity: Kendra Martin, Kathryn Bassett, Elham Ali

dundurn.com   dundurnpress
@dundurnpress   dundurnpress
dundurnpress   info@dundurn.com

FIND US ON NETGALLEY & GOODREADS TOO!

DUNDURN